Hyung Je Jo · Jun Ho Jeong · Chulsik Kim

Agile Against Lean

An Inquiry into the Production System of Hyundai Motor

palgrave
macmillan

Hyung Je Jo
University of Ulsan
Ulsan, Korea (Republic of)

Jun Ho Jeong
Kangwon National University
Gangwon-do, Korea (Republic of)

Chulsik Kim
The Academy of Korean Studies
Gyeonggi-do, Korea (Republic of)

ISBN 978-981-99-2041-9 ISBN 978-981-99-2042-6 (eBook)
https://doi.org/10.1007/978-981-99-2042-6

This Palgrave Macmillan imprint is published by the registered company Springer Nature Singapore Pte Ltd.
The registered company address is: 152 Beach Road, #21-01/04 Gateway East, Singapore 189721, Singapore

Preface

In 1976, Hyundai Motor Company (Hyundai Motor hereafter) released the Pony, a passenger car shaped like 'a rooster with a missing tail.' The company which had previously assembled and produced Ford's Cortina with a knockdown (KD) had now unveiled its own first model. The Pony was the sixteenth own model to be released globally and the second in Asia, following on from Japan. At the time, a Pony cost 2.3 million Korean won, and Hyundai Motor's annual production capacity was only 56,000 units. Although domestic consumers were enthusiastic about the Pony, which featured a unique design and decent performance, automobile experts were watching the story of Hyundai Motor as a latecomer with mixed eyes.

Half a century later in 2022, Hyundai Motor launched the Sports Utility Vehicle (SUV) Ioniq 5, the company's first electric vehicle model. The basic design of this car pays homage to the Pony. The Ioniq 5 won 'World Vehicle of the Year' at the 2022 World Car Awards, one of the world's top three automobile awards, and received favorable reviews from consumers and experts around the world in terms of not only design but also driving and performance. The basic model of the Ioniq 5 sells for 50 million Won, and the annual production capacity of Hyundai Motor Group, including Hyundai Motor and Kia Motors (hereinafter referred to as Kia), is 9.14 million units. Hyundai Motor has now developed from an automaker that sought to catch up with advanced carmakers to one that has overtaken them. In 2022, Hyundai Motor ranks third in the world after Toyota and Volkswagen in terms of sales volume, and second in electric cars after Tesla.

The past 50 years have been a time of compressed growth for Hyundai Motor that mirrors the development of the Korean economy. Korea was no different from many impoverished African nations in the early days of industrialization, but has now transformed into an advanced country that exports semiconductors and electric vehicles. This book aims to show that the driving force behind this transformation results from the competitiveness of Korea's flagship manufacturing industry. As the late Alice Amsden explained in her famous book *Asia's Next Giant* (1989), the success of Hyundai Motor, a typical 'national champion manufacturer' in Korea, can be summarized as a shift from imitation to innovation. Unlike other latecomers, Hyundai Motor's organizational capabilities have leapt from the level of imitating more advanced competitors to the stage of innovating in their own right.

What made this leapfrogging possible? There are certainly core elements such as the firm's status as a *chaebol*, as well as the engineers, labor–management relations, and inter-firm relations that have allowed the manufacturing industry to achieve such compressed growth in the Korean economy. These elements have also been highly condensed in the automobile industry. In the same sense that Toyota represents the Japanese economy, Hyundai Motor represents Korea. This book aims to explain the secret of Hyundai Motor in developing from a latecomer that relied on foreign technology to make low-end cars to an advanced carmaker that boasts proprietary technology and makes luxury cars that attract attention around the globe. At the same time, unlike existing literature in this field, we seek to shed light on the dark side of this success. Korea's automobile industry is highly competitive but also suffers from the side effects of labor market segmentation and widening gaps between companies.

We agreed to write this book because we would like to explain this reality from our own perspective. The simple application of Western theoretical discourses is unable to explain the compressed growth of the Korean economy. With the same logic, we oppose the view that Hyundai Motor's production system is an epigone or variant of lean production systems such as Toyota's. In order to explain Hyundai Motor's story, we need a dialectical interaction between theory and reality rather than the unilateral adoption and application of a Western theoretical framework. The genuine character of Korea's economic development and Hyundai

Motor's remarkable growth can only be elucidated through a deep exploration involving empirical research and a critical mindset grounded in reality. This book is the result of a collective study to theoretically and empirically understand Hyundai Motor's compressed growth. One of the authors, Hyung Je Jo, has already made a preliminary presentation on the concept of 'agile production systems' in 2016. This book elaborates on this critical problématique in more detail. We have had to repeat the dynamic work of applying various theoretical resources to the case of Hyundai Motor and reconstructing this theoretical framework based on reality. We attempt to systematically present the theoretical and empirical characteristics of Hyundai Motor's agile production system.

This book consists of three parts. In Part I, we explain 'authoritative experimentalism,' a theoretical concept that integrates Hyundai Motor's compressed growth, with a focus on the three keywords catch-up, *chaebol*, and production system in Chapter 1. In Chapter 2, we summarize the major characteristics of Hyundai Motor's 'agile production system' through historical and empirical analysis, mainly capitalizing upon the audit reports of companies.

Part II examines the elements that make up Hyundai Motor's agile production system through four chapters. We explain the skill formation and *ad hoc* or improvisational characteristics of engineers that enabled Hyundai Motor's innovation in Chapter 3. In Chapter 4, we analyze the flexible automation and skill-saving work organization that made this possible. Chapter 5 describes the extended quasi-vertical integration of inter-firm parts suppliers which expands across not only Hyundai Motor's affiliates but also legally independent and external parts suppliers. Finally, Chapter 6 deals with Hyundai Motor's overseas green-field plants where the agile production system has been transferred to new locations. These four aspects work organically with one another through interrelationships.

Part III provides a summary and an examination of the duality of Hyundai Motor's agile production system, with future prospects presented in Chapter 7. Behind Hyundai Motor's compressed growth lies a deepening segmentation in the Korean labor market and a widening gap between companies. For example, the segmentation of 'regular' and 'non-regular' workers and the widening gap between large *chaebol* firms and small and medium-sized enterprises (SMEs) are inevitable side effects that Hyundai Motor has borne witness to in the process of compressed growth. Nowadays, the company faces a new challenge in the industrial transformation to eco-friendly and electric vehicles. It remains to be seen whether Hyundai Motor's production system will follow a 'path-dependent' or 'path-breaking' course in the future. At the same time, this is also a question about the future of the Korean economy in general.

Ulsan and Seoul, Korea

Hyung Je Jo
Jun Ho Jeong
Chulsik Kim

Acknowledgments

In publishing this book, we cannot fail to mention the many people to whom we owe our thanks. Above all, we express our sincere gratitude to the former and current engineers, managers, and employees of Hyundai Motor Group and parts manufacturers who understood the need for academic research and assisted us in spite of being busy with work at local and overseas plants. We especially thank Dong Woo Choi, Bu Young Ha, Eon-Tae Ha, Kimyoung Hwang, Jin Taek Kim, Koon Rack Lee, Seong Gyu Lee, Inho Park, Sang Hyuck Park, Jang Won Seo (in alphabetic order). We regret that we are unable to name you all individually, but it is clear that this book could not have made it to print without your cooperation. Nevertheless, it should be noted that all responsibility for the details of this book lies with the authors themselves.

We would also like to thank our colleagues for understanding the theoretical and practical meaning of our research in the harsh reality of academia, and for your unsparing support and encouragement. In particular, we would like to give special thanks to Moon-Ho Rhee, Byeong Cheon Lee, Jong-Seong You, Eehwan Jung, Byung-Hoon Lee, Jewheon Oh, Seong-Jae Cho, Jang-Pyo Hong, Jeong Hee Lee, Joongsan Oh, Ah Rom Kang, the late Michel Freyssenet, Ulrich Uergens, Martin Krzywdzinski, Thomas Heipeter, and A. J. Jacobs. Moreover, we have been indebted to Byeong Cheon Lee for inspiring the idea of 'authoritarian experimentalism' during the lively conversation. With your support and encouragement, we have been able to develop our critical problématique and make up for our own lack of expertise. However, we emphasize that we are solely responsible for any errors or mistakes in this book.

Finally, we would like to thank Palgrave Macmillan for recognizing the value of this book and deciding to support and publish it. We are delighted that the findings of our research will be made available to academia and industry through the publication of this book. We are grateful to the editor of *Journal of Asian Sociology* for allowing us to use the following articles published in the journal in writing Chapters 3 and 5 of this book, respectively: Kim, Chulsik, Jun Ho Jeong, and Hyung Je Jo. 2021. Detecting Dynamic Changes in Hyundai Motor's Parts Supply System as an Industry Latecomer: The Quasi-Vertical Integration of Internal and External Networks, *Journal of Asian Sociology*, 50(1), 55–89; Kim, Chulsik, Jun Ho Jeong, and Hyung Je Jo. 2022. Improvisational Intensive Problem-Solving Capability: The Case of Hyundai Motor's New Car Projects, *Journal of Asian Sociology*, 51(4), 323–354.

Contents

Part I Conceptual Blocks and Overview

1 Introduction—Building Conceptual Blocks for an Understanding of Catching Up, *Chaebols*, and Production Systems 3
1.1 Introduction 3
1.2 The Conceptual Blocks of Catching Up, Chaebols, *and Production Systems* 5
1.2.1 The Catch-Up of the Korean Economy 5
1.2.2 Catching Up and Forms of Business Organization 12
1.2.3 Automotive Production Systems 35
1.3 Conceptualizing Hyundai Motor's Agile Production System as Authoritarian Experimentalism 43
1.3.1 Conceptual Map of This Book 43
1.3.2 Hyundai Motor's Agile Production System: An Authoritarian Experimentalism 46
References 53

2 An Overview of Hyundai Motor's Production System 65
2.1 Hyundai Motor's Production System Before the 1997 Asian Financial Crisis 66
2.2 A Macro-institutional Overview of Hyundai Motor's Production System After the 1997 Asian Financial Crisis 70
2.2.1 The 'Middle-Ranked Carmaker's Trap' 70

2.2.2 *Entrepreneurial Management Exploits 'Windows of Opportunity'* 73
2.2.3 *Corporate Governance Structure and the* Chaebol *System* 77
2.2.4 *Product Strategy and Profitability—The Pursuit of Flexible Scale-Diversity* 86
2.3 *A Micro Overview of Hyundai Motor's Production System After the 1997 Asian Financial Crisis* 96
2.3.1 *Skill-Building in Engineers: Learning by Improvisation* 96
2.3.2 *Work Organization and Shop-Floor Workers: Flexible Automation, Skill-Saving, and Segmented Labor* 102
2.3.3 *Inter-Firm Relationships: Extended Quasi-Vertical Integration* 108
2.3.4 *Overseas Transplant Strategy: The Geographic Transfer of Hyundai Motor's Production System* 111
References 112

Part II Empirical Anatomy

3 Skill Formation in Engineers—Learning by Improvisation 119
3.1 *Introduction* 119
3.2 *Crisis in the Business Environment and Construction of the Pilot Center* 122
3.3 *New Car Development and Intensive Problem-Solving Capacity* 125
3.3.1 *The Collective Skills of Engineers* 125
3.3.2 *The New Car Development Process and Intensive Problem-Solving* 129
3.3.3 *Comparison with Toyota* 135
3.4 *The Vertical Restructuring of Production Engineering* 137
3.4.1 *Horizontal Production Engineering at Hyundai Motor* 137
3.4.2 *Integrated Production Engineering in Japan* 140
3.4.3 *The Establishment of Hyundai Motor's Agile Production System* 141
3.5 *Conclusion* 143
References 144

4 Work Organization and Shop-Floor Workers—Flexible Automation, Skill-Saving, and Segmented Labor 147
4.1 Introduction 147
4.2 Background to Skill-Saving Work Organization—Before the 1997 Financial Crisis 149
4.2.1 Development of Production Technology and Labor-Replacing Automation 149
4.2.2 A Failed Attempt to Build Workplace Skills 150
4.3 Full-Scale Flexible Automation and Skill-Saving Work Organization in the 2000s 152
4.3.1 Development of Modular Production 152
4.3.2 Flexible Automation and Informatization 157
4.3.3 Skill-Saving Work Organization 158
4.3.4 Multi-Tiered Quality Control 161
4.3.5 Skill-Saving Work Organization, Labor–Management Relations, and Labor Market Segmentation 162
4.3.6 Comparison with Toyota Production System 165
4.4 Conclusion 166
References 168

5 Supplier Relations—Extended Quasi-Vertical Integration 171
5.1 Introduction 171
5.2 Formation of Extended Quasi-Vertical Integration—Before the 1997 Financial Crisis 173
5.2.1 The Formation of the Subcontracting System 173
5.2.2 Captive Relationships and Paternalistic Control 175
5.3 Systematization of Extended Quasi-Vertical Integration in the 2000s 176
5.3.1 Modular Production and Fostering Affiliate Modular Makers 177
5.3.2 Modular Production and Reorganization of the Parts Supply Structure 181
5.3.3 Comparison with Toyota 185
5.3.4 Quality Management and Systematic and Bureaucratic Control 187
5.4 Factors in the Formation of the Extended Quasi-Vertical Integration 189

5.5 *Japan's* Keiretsu *Network and Hyundai Motor's Extended Vertical Quasi-Integration* 191
5.6 *Conclusion* 193
References 194

6 The Overseas Production Network—'Model Factories' and the Global Transfer of Hyundai Motor's Production System 197
6.1 *Introduction* 197
6.2 *Transplanting the Agile Production System Around the World* 199
6.2.1 *The 'Model Factory' Concept and Construction of Green Field Plants* 200
6.2.2 *The Application and Adaptation of Human Resource Management* 202
6.2.3 *'Follow Sourcing' by Korean Parts Suppliers* 205
6.3 *Operating Organizational Capabilities on a Global Scale* 207
6.3.1 *Concentrating New Car Development at the Headquarters* 208
6.3.2 *Daily Plant Operations* 210
6.3.3 *Human Resource Management for Locally Hired Staff* 212
6.3.4 *The Parts Supply System* 215
6.3.5 *Comparison Between Hyundai Motor and Japanese Automakers* 217
6.4 *Conclusion* 219
References 221

Part III Reflections and Prospects

7 Conclusions—Theoretical and Empirical Reflections and Future Prospects 225
7.1 *Theoretical Contributions* 225
7.1.1 *The Catch-Up Process* 226
7.1.2 *The* Chaebol *System* 228
7.1.3 *The Automotive Production System* 230
7.2 *Empirical Findings* 233
7.2.1 *Skill-Building in Engineers* 234
7.2.2 *Work Organization and Shop-Floor Workers* 235

7.2.3 *Inter-Firm Relationships* 237
7.2.4 *Overseas Transplants* 237
7.2.5 *Comparison Between Hyundai Motor and Toyota's Production Systems* 238
7.3 *Two Sides of Hyundai Motor's Production System and Future Prospects* 243
7.3.1 *Two Sides of Hyundai Motor's Production System* 243
7.3.2 *Future Prospects* 246
References 250

Index 255

About the Authors

Hyung Je Jo is an emeritus professor of social science at the University of Ulsan, Ulsan, South Korea. He has a Ph.D. in sociology from Seoul National University, Seoul, South Korea. His research interests are industry, region, production systems, and labor relations. He has published numerous papers and books on the Korean automotive industry for three decades. His latest papers are a comparative study between German and Korean automakers operating in Central and Eastern Europe published in *Critical Perspectives on International Business*, *Competition & Change*, and *Journal of Asian Sociology*. Currently he is working on the employment change in the transition to eco-friendly automobile industry.

Jun Ho Jeong is a professor in the College of Social Sciences at Kangwon National University, Chuncheon, South Korea. He has a D.Phil. in economic geography from the University of Oxford, Oxford, United Kingdom. His research interests include national and regional development issues, industrial and innovation policies, and the dynamics of income and wealth inequalities. Previously, he worked at the Korea Institute for Industrial Economics and Trade, a government-funded think-tank, as a research associate and the director of the institute's economic survey and forecasting division, and served as an economic advisor to the Prime Minister. Currently he is working on changes in the Korean production system at the national and regional levels in relation to the changing geopolitical and economic circumstances.

Chulsik Kim is an assistant professor in the division of social science at the Academy of Korean Studies, Seongnam, South Korea. He has a Ph.D. in sociology from Seoul National University, Seoul, South Korea. His research interests are industry, organization, labor market, and employment relations. He has written a doctoral thesis on the modularization of Hyundai Motor from the perspective of the value chains, and his latest papers have been published in *Journal of Asian Sociology.* Currently he is working on the relationship between technology and work in the digital age.

List of Figures

Fig. 1.1	Conceptual map of this book	44
Fig. 2.1	Hyundai Motor's profitability and relative quality level in the 1980s and 1990s	72
Fig. 2.2	Governance structure of *Chaebols*	79
Fig. 2.3	Hyundai Motor Group's subsidiaries and employment by industry	80
Fig. 2.4	Hyundai Motor Group's equity structure	81
Fig. 2.5	Roundabout shareholding structure with multi-cores (%): Hyundai Motor Group	83
Fig. 2.6	Quasi-internal transactions of Hyundai Motor Group: Hyundai Motor's claim-obligation relationships with affiliates in relative terms	85
Fig. 2.7	Hyundai Motor's profitability (%)	88
Fig. 2.8	Comparison of production and relative stock price between Hyundai Motor and Toyota	90
Fig. 2.9	Comparison of several automakers and auto parts companies among BCG's 50 most innovative companies	91
Fig. 2.10	Correlation between Hyundai Motor's markup and exchange rate	91
Fig. 2.11	Product composition of major automakers (%)	93
Fig. 2.12	Changes in relative quality index of Hyundai Motor and Toyota	94
Fig. 2.13	International comparison of engineers in the automotive industry (%)	97

Fig. 2.14	Comparison of the number of active patents of major automakers	98
Fig. 2.15	Comparison of R&D investment to sales ratio between Hyundai Motor and Toyota (%)	99
Fig. 2.16	Dimensions of organizational improvisation	101
Fig. 2.17	Hyundai Motor's fixed capital investment compared to cash flow	104
Fig. 2.18	Hyundai Motor's investment propensity and robot density in the automotive industry	105
Fig. 2.19	Trends in Hyundai Motor's labor share (%)	108
Fig. 2.20	Changes in Hyundai Motor's labor and material costs (%)	109
Fig. 2.21	Extended Quasi-Vertical Integration in the Korean automotive industry	110
Fig. 2.22	Comparison of overseas production between Hyundai Motor Group and Toyota Group (%)	112
Fig. 3.1	Hyundai Motor's new car development process	130
Fig. 3.2	Organizational chart of the cross-functional team at the Hyundai Motor's pilot center	132
Fig. 3.3	Changes in Hyundai Motor's organizational capabilities in production engineering	143
Fig. 4.1	Map of ERP concept	158
Fig. 4.2	Final assembly line layout of Hyundai Motor's Ulsan plant	160
Fig. 5.1	Number of first-tier parts makers for Hyundai Motor	183
Fig. 5.2	Proportion of automobile parts SMEs in each transaction layer	184
Fig. 6.1	Preparatory procedures for mass production at Hyundai Motor's overseas plants	209
Fig. 6.2	The transfer of Hyundai Motor's production system to overseas plants	216

List of Tables

Table 1.1	Comparison of Iacocca Institute Production Systems	47
Table 2.1	Ownership structure of the five largest *Chaebols* (as of the end of 2021)	82
Table 2.2	Hyundai Motor-Kia Motors platform integration	92
Table 2.3	Number of new products (passenger vehicles) by Hyundai Motor and Kia Motors	92
Table 3.1	Dimensions of Hyundai Motor's improvisational product development model	128
Table 3.2	A comparison of the new car development process: Hyundai Motor vs. Toyota	138
Table 4.1	Hyundai Motor's automation by plant	151
Table 4.2	Hyundai Motor's modularization	155
Table 4.3	A comparison of Toyota's production system and Hyundai Motor's	166
Table 5.1	Suppliers' council membership by parent company (as of 1991)	174
Table 5.2	Number of parent companies for first-tier parts suppliers (as of 1997)	176
Table 5.3	Sales and profits by business division at Hyundai Mobis (Unit: KRW billion, %)	180
Table 5.4	Hyundai Mobis' M&As in the modular parts business (2000–2007)	182
Table 5.5	Trends in operating profit margins of Hyundai Motor's parts suppliers (%)	184
Table 6.1	Production capacity of Hyundai Motor Group (as of 2022)	198

Table 6.2 Employees at Kia Motors Slovakia in 2015 203
Table 6.3 Korean automotive parts suppliers responsible for follow sourcing in Central and Eastern Europe 206
Table 6.4 Characteristics of production processes at Kia Motors Slovakia 211
Table 6.5 Local procurement rate of Korean auto parts makers responsible for follow sourcing in Europe (as a percentage of total purchase amount) 215
Table 6.6 Transferability of organizational capabilities in Japanese and Hyundai Motor Group's overseas automotive plants 220
Table 7.1 Comparison of Hyundai motor and Toyota's production systems 240
Table 7.2 Two models for future vehicle transformation in the automotive industry 249

PART I

Conceptual Blocks and Overview

CHAPTER 1

Introduction—Building Conceptual Blocks for an Understanding of Catching Up, *Chaebols*, and Production Systems

1.1 Introduction

The automobile industry has always been technology and capital-intensive, requiring a substantial demand base to be profitable. As a result, for a long time it was the preserve of developed countries. From the 1960s to 1980s, the global market consisted of an oligopoly structure centered around the 'Big 3' in the US, five major automakers in Japan and six major carmakers in Western Europe, as well as some luxury brands such as Mercedes-Benz and BMW. Armed with a lean production system, Toyota then began to make rapid progress from the 1980s and became a new standard in the automobile industry, with American companies struggling to survive. However, there was also one enterprise from a developing country that emerged on the scene and began to rapidly break through the market structure of the automobile industry to become one of the top five global players in the 2010s. That company is Korea's Hyundai Motor Company (hereafter Hyundai Motor).

The technological standards of the automobile industry were established in the 1920s, and the automatic transmission has been the only fundamental technological innovation in automobile structure since that time (Abernathy, 1978). With the introduction of electronic control devices, the technological innovation in the automobile industry has been referred to as 'rapid gradual innovation' (Clark & Fujimotto, 1991; Fujomoto, 2014). This has traditionally made it difficult for newcomers in

H. J. Jo et al., *Agile Against Lean*,
https://doi.org/10.1007/978-981-99-2042-6_1

developing countries to catch up with major players in developed nations. Against this background, Hyundai Motor has gone through a difficult process of initial learning from Mitsubishi before eventually establishing its own production system to stand shoulder to shoulder with global carmakers.

Since World War II many developing countries have tried to foster growth in the automobile industry as a strategic means of promoting economic development, but Hyundai Motor was virtually the only successful example of such a policy until China appeared. Hyundai Motor's remarkable growth since the 1997 Asian Financial Crisis was made possible through the establishment of its own production system. While advanced companies such as Toyota have stagnated after the 2008 Global Financial Crisis, Hyundai Motor has flourished and taken steps toward leading the global automobile market. Hyundai Motor's production system has a number of unique characteristics that will be discussed in the following chapters. Hyundai Motor has become competitive on a global level through modularizing the supply of auto parts, pragmatic experiments led by engineers, and bold overseas investment in line with the development of the global value chain.

This book mostly focuses on the period in which Hyundai Motor overhauled its production system after the Asian Financial Crisis and then made great strides forward in the 2010s. Key concepts from this period include catching up, the Korean corporate governance structure known as a *chaebol*, and production systems. This chapter examines and summarizes prior research and discussions on these keywords. These theoretical and empirical discussions provide a framework for critically examining the amazing achievements of Hyundai Motor.

The questions raised in this book include the following. As a latecomer from a developing country, what factors made it possible for Hyundai Motor to catch up to more advanced competitors? What role has the corporate governance structure of the *chaebol* played in making it possible for Hyundai Motor to compete with carmakers in more developed countries? What are the characteristics of Hyundai Motor's production system in comparison to the industry standard set by Toyota? And is Hyundai Motor's remarkable growth temporary or sustainable?

1.2 The Conceptual Blocks of Catching Up, *Chaebols*, and Production Systems

This section begins with a discussion of three key concepts required to provide the theoretical context for the structure of this book. The formation of Hyundai Motor's own production system, a central part of this book, has been a continual process of 'catching up.' As the existing research in all three fields is quite extensive, we focus our discussion in relation to the structure of this book and address key concepts involving the production system and corporate governance of a catching-up automaker.

1.2.1 The Catch-Up of the Korean Economy

1.2.1.1 The Catch-Up Mechanism of the Korean Economy

Catching up refers to narrowing the *per capita* income and productivity gaps between latecomers and forerunners in an economic sense (Fagerberg & Godinho, 2005). It is also about understanding the relationship between technological innovation and economic growth. For example, latecomers such as the United States, Germany, and Japan compared to the United Kingdom have achieved success through organizational innovations including mass production and economies of scale, R&D innovation in the chemical and engineering industries, and the just-in-time production system, respectively.

European examples (e.g. Abramovitz, 1986; Gerschenkron, 1962) and discussions of developing countries in East Asia (e.g. Amsden, 1989; Johnson, 1982; Kim, 1997; Kim & Nelson, 2000; Wade, 1990) are related to the concept of 'capability building,' Catch-up is not automatic, but a difficult process that requires great effort and capacity. Knowledge and skills are created through learning or organized R&D, and are not easily transferable because they are deeply embedded in the specific capabilities of private companies and their networks. Gerschenkron (1962) raised the need for capacity building to successfully catch up, and focused on investment banking as a new 'institutional means' of mobilizing resources in the sense that capital and technology-intensive technologies were widely utilized when Germany caught up with Britain.[1] Gerschenkron believed that transferring technology was not easy and that active government intervention in the market was a necessary ingredient. On the other hand, Veblen (1915a) argued that progress in 'machine-technology' facilitated the transfer of technology between countries, and as a result, catching up depended on pecuniary factors and opportunities in new industries. Veblen believed that the transfer of technology was not difficult in terms of skills and infrastructure because the technology was readily available, and the market could fulfill the necessary coordination role in the absence of active intervention from outside institutions.

There are largely two positions with regard to the catch-up of East Asia. One is the argument of the World Bank (1993) which explains the economic miracle of East Asian countries via a market-friendly catch-up process and the ensuing shared growth, which is reminiscent of Veblen's (1915a) idea. The other emphasizes the intervention of strong governments as pivotal in the successful catch-up (e.g. Amsden, 1989; Johnson, 1982; Kim, 1997; Wade, 1990), which is implicitly or explicitly based on Gerschenkron's (1962) argument.

The latter is an important part of technological capability building and developmental state theories. Kim (1980, 1997) established the technological capability building theory on technological catch-up through the lens of Korea. He divided the technological development stages of latecomers into 'replicative imitation, creative imitation, and innovation,' demonstrating the processes through which latecomers acquire and learn advanced technologies. These channels include the import of capital goods, technology licensing, reverse engineering, trade policies, and less strict intellectual property rights standards. However, Kim's discussion is based on the perception of technology as cumulative and linear (Pérez & Soete, 1988) using the framework of the reverse product life cycle, as just mentioned, unlike the standard one in developed countries. However, since catching up as path creation or path-skipping is only possible when technological innovation is understood as a non-linear process (Lee & Lim, 2001), it is virtually impossible for developing nations to catch up to or overtake advanced countries under this sequential approach. In

[1] In addition, institutional means for catching up include linkage between advanced technologies and markets (and users), supply of necessary skills, services, and other input elements, and regional innovation systems and networks (Verspagen, 2005: 536).

addition, Kim was only interested in the formation of R&D and engineering capabilities and did not mention anything about skill formation on a workplace level.

Developmental state theory believes that active state intervention in the market is crucial, justified by economic growth (Johnson, 1999). Johnson (1982, 1999) identifies Germany, Japan, and Korea as the same type of developmental state.[2] He classifies these countries as rationally planned economies based on the synergy between markets and planning and contrasts this with commanding states where planning replaces the market (such as the former socialist countries) and market-coordinating state or market rational states (such as the United States). Evans (1995) explains that the Korean government built a relationship of public–private partnerships and embedded autonomy to nurture *chaebols* on the basis of bureaucratic policy capacity. Amsden (1989) argues that Korea succeeded in industrializing as a latecomer thanks to the state's performance disciplines which linked preferential support for *chaebols* with economic indicators such as export performance and investment in strategic industries, and notes that Korea emerged as 'Asia's next giant' after Japan. As mentioned above, the system building of reciprocity in which performance discipline was linked to preferential support was a decisive element in the success of Korea's industrialization. Aoki et al. (1997) conceptualized preferential support from the state as 'contingent rent,' with the state's conditional enforcement of performance discipline.

The authoritarian Korean government made social agendas, such as labor and welfare, subordinate economic growth by utilizing depoliticization, strong dictatorship, anti-communist rhetoric, and the Cold War regime to ensure the smooth operation of this contingent rent system. This acted as an institutional compulsion for the creation and distribution of contingent rent (Kahn & Blankenburg, 2009).[3] In order to solve the problem of investment coordination, policies including financial compression and the redistribution of wealth through forced savings and inflation

[2] Harvey (2003) also pointed out commonalities between Germany and East Asian developmental states in that the strong intervention of state power contributed to capital accumulation. Accordingly, Germany's Bismarck, Japan's Meiji Restoration, and Korea's Park, Chung Hee regime all fall under the common tradition of developmentalism.

[3] Khan (2000) argues that Korea's industrialization is a special case that is not easy to apply to other countries because it was based on depoliticization excluding civil society and labor, intensive investment and control of resources by the state, and technology learning.

were implemented, and some land and property rights were also reserved and restricted for the sake of economic growth in Korea.

However, discussions such as Amsden (1989) and Aoki et al. (1997) fail to appreciate that the compressed growth of *chaebols* through the creation of contingent preferential rent could cause changes in power and class structure and bring serious difficulties to the state, market, or civic discipline of *chaebols* on the path of post-catch-up development after Korea became a democratic nation. This has been called the 'paradox of democratization' (Lee, 2020). In a similar vein, Selwyn (2011) criticizes Gerschenkron's (1962) discussion for not integrating the role of political processes and social class into the theoretical framework.

As noted above, the developmental state seeks legitimacy in economic achievements (Johnson, 1982, 1999). However, the developmental state that comes 'after' an initial developmental state faces the dual challenges of upgrading production capability and achieving greater social inclusion (Whittaker et al., 2020). In this way, the concept of an 'adaptive' developmental state has been proposed to capture the ongoing transition process of developmental states (Wong, 2004). Regarding the concept of the 'adaptability' of developmental states and 'embedded autonomy' between the state and civil society (Evans, 1995), there have been some reflections on inclusive and sustainable growth and the challenge of integrating both economic and social progress (see Chang & Andreoni, 2020; Evans, 2021; Wade, 2018; Whittaker et al., 2020).

For a country to successfully catch up, the state must have the ability to share a vision, coordinate diverse interests, and mediate social conflicts. This is called a 'hard state.' This state can be either authoritarian or democratic (Unger, 2009, 2019). To date, the developmental state theory on the East Asian experience mainly leans toward the former (Myrdal, 1968). However, if success does not depend on political authoritarianism, developmental governance capacity should be distinguished from the authoritarianism (Evans, 1995). A developmental hard state may have the capacity to manage and discipline capital (Amsden, 1989). However, if this state's disciplinary capacity is loosened without the spread of civic democracy throughout society, it could lead to soft market liberalization or the involution of developmental states degenerating into crony capitalism (Aoki, 2010). The previous discussion of developmental state theory misses these points.

1.2.1.2 Major Empirical Studies on the Korean Economy's Catch-Up Process

Thus far, we have looked at prior literature on the mechanisms behind Korea's catch-up process. This section examines some major empirical studies on the characteristics of this process. Hattori (2005) proposed the concepts of 'processing type' and 'assembly type' industrialization. According to Hattori, Japan is a case of the former while Korea belongs to the latter category. The assembling-type explicitly or implicitly assumes a production system in which machines are able to easily substitute for or exclude labor and a dual structure between large corporations and small and medium-sized enterprises (SMEs). This helps to explain Korea's chronic trade deficit with Japan in the fields of machinery and equipment, materials, and parts. Hattori's argument is that an industrial base for assembling was built in Korea, with core parts, materials, and mechanical equipment imported, through a combination of factors, namely the increased supply and lower price of numerical controller (NC) automation machinery in the 1970s, the international division of labor following the Cold War, and the Korean government's active industrial policy. The 'assembly type' means that if a Korean company wanted to produce a high-tech product, the processing machinery, parts, and materials had to be imported from other countries, especially Japan, since it was impossible to localize these resources in a short period of time. This theory implies an industrialization pattern in which the accumulation of technology is slow under geo-economic technological conditions similar to those of Japan, although it is also reminiscent of the relative 'easy' catch-up effect under the technical conditions mentioned by Veblen (1915a).

Levy and Kuo (1991) argue that the assembly strategy based on subcontracting between large companies and SMEs has played an effective role in Korea's growth. This strategy involves beginning operations even if the firm faces unit costs that exceed the market price. The goal is to acquire design and operating capabilities with accumulating technology experience inside the company through economies of scale and learning by doing. As is well known, this strategy requires massive upfront capital investment and an industrial structure that is centered around large corporations. In addition, since this strategy requires a technology learning process that moves up from simple to complex technology, acquiring the latest process and production design technology is important for improving productivity and becoming competitive. Accordingly, this requires promoting skill-building for engineers in charge and motivating them to work hard.

Fujimoto (2006) points out that Korean companies have a competitive advantage in capital-intensive open modular products such as semiconductors, as well as general-purpose steel and petrochemical products. He argues that the sectors in which Korean conglomerates have rapidly caught up with or surpassed incumbents are those in which they were able to compete through product upgrades, rather than by acquiring complicated workplace skills or technical capabilities. Fujimoto also suggests that the reason Korean conglomerates remain competitive in this sector is the outstanding ability of *chaebols* to mobilize enormous amounts of financial resources and make prompt decisions through the concentrated organizational structure.

Lee (2013) argues that specialization in technologies with short technology life cycles (such as the IT sector [Park & Lee, 2006]) has been an effective strategy in Korea's transition from a middle-income country to a developed nation. According to Lee, although (reverse) product life cycle theory is relevant to the initial catch-up process, the technology life cycle theory based on a technology's degree of longevity is more appropriate when discussing the transition from middle-income to high-income countries. Korea's catch-up can be understood as the process of transitioning from technology with a long life cycle to technology with a short life cycle. For example, specialization in Korea has come about through sectors with long, low value-added technology life cycles (e.g.: apparel) in the 1960s, medium technology life cycles (e.g.: automobiles and shipbuilding) since the mid-1970s, and short technology life cycles (e.g.: digital electronic switching systems, semiconductors, mobile phones, and digital TVs) since the mid-1980s. According to Lee and Lim (2001), although catching up can be classified into the 'path following,' 'path-skipping,' and 'path-creating' types, successfully catching up requires only the latter two cases. This is the result of utilizing 'windows of opportunity' such as the emergence of new technologies (e.g. digital technology), economic fluctuations or changes in market demand (e.g. recession), and government intervention or regulatory changes.

The literature contains a discussion of the 'middle-income trap' in which developing countries fail to transition from middle-income countries to developed nations. The World Bank warned in 2007 that developing countries in Asia have fallen into this trap (Gill & Kharas, 2007).

This suggests that the transition from a middle-income to a high-income country is qualitatively different from the transition from a low-income to a middle-income country (Aghion et al., 2021). Therefore, in order for the former to work, it is necessary to completely overhaul the institutions, structures, and policies that have been aligned with the latter. Acemoglu et al. (2006) note that an investment-based growth model is feasible for the transition from a low-income to a middle-income nation. In this strategy, existing technologies are imported through significant capital investment and intensive intervention from the state or business groups such as Korean conglomerates. However, in order to avoid falling into the middle-income trap, countries must switch to an innovation-based growth model that can accommodate the invention of new technologies in the traditional growth model by creating a dynamic competitive environment. This involves training skilled labor, R&D investment in advanced technologies, promoting competition between companies, and mitigating barriers to entry and exit.

Based on these discussions, Aghion et al. (2021) suggest that Korea is a case of a successful transition from an investment-based growth model to an innovation-based growth model due to open and competition-oriented reform policies after the 1997 Asian Financial Crisis. This study notes that the total factor productivity (TFP) of *chaebol* companies had stagnated or declined before the 1998 reforms, but subsequently increased rapidly for both *chaebol* and non-*chaebol* companies. The TFP's growth was remarkable in industries already dominated by *chaebols*, which were heavily affected by the reforms, and productivity growth in non-*chaebol* companies also saw a sharp uptick. In addition, numerous non-*chaebol* companies entered all industries in the wake of the reforms, and the increase in patent activity was more prominent among non-*chaebols*, representing a turnaround from the pre-reform period.

In summary, the literature notes that Korea's successful catch-up could be attributed to large-scale facility investment to enjoy economies of scale, greater emphasis on production design and process technologies with less attention paid to workplace skills, standardization, *chaebol* governance, and the government's industrial policies as a means of sharing risk. In other words, the combination of the effect of Veblen (1915a)'s machine-technology and the institutional approach of Gerschenkron (1962) provides a useful and relevant explanation for Korea's successful industrialization strategy.

This means the remaining questions are investment and market demand issues. The *chaebol* system and government policies are institutional apparatuses to solve these problems. For example, steel, shipbuilding, semiconductors, mobile phones, and digital TVs have become competitive thanks to timely investment by *chaebols* through standardization and modularization and financial support from the Korean government. When R&D and manufacturing design capabilities are compared with shop-floor skills, it is clear that the internal accumulation of engineering capabilities played a more important role in catching up, as described by Amsden (1989).

On the other hand, as analyzed in Aghion et al. (2021), The TFP has increased in both *chaebol* and non-*chaebol* firms in industries that have been heavily affected by open and competition-oriented reform policies since the 1997 Asian financial crisis (e.g.: automobiles). This suggests that both firms have improved their innovation capabilities and reorganized their cooperation between them, for example, through the modularization of parts since the 1997 Asian Financial Crisis in the automotive industry.

1.2.2 Catching Up and Forms of Business Organization

In this section, we focus on corporate organizational forms and examine the catch-up process from a theoretical level. It is generally accepted that there are three ways for latecomers to pursue industrialization to catch up (Schmitz, 1999: 478). The first is led by SMEs, the second involves attracting large external companies or developing such companies in a planned manner, and the third is a combination of the two. Industrialization is aimed at stimulating economic growth. This is linked to the creation of a base for increasing returns derived from economies of scale and scope, learning effects, and network effects (Shiozawa et al., 1997).

Marshall (1890) distinguished between internal and external economies. The former refers to savings in production costs applied within firms, while the latter is between firms. From the perspective of business structure, the combination of economies of scale (and scope) and internal and external economies as the fundamental basis for increasing returns has historically led to different worlds of corporate organization. The first is external Marshallian economies of scale centered around SMEs and the second is internal Schumpeterian economies of scale based on large corporations. There is also a hybrid model that incorporates aspects of both. In these three worlds, learning and network effects mesh across different geographic, historical, and institutional contexts.

1.2.2.1 The World of Marshallian SMEs

The first type of industrialization is called 'industrialization from below.' This is an association of small and medium-sized enterprises that aims to build an associational economy, and typical examples include the industrial districts of England in the nineteenth century and the Third Italy. Although Sabel and Zeitlin (1997) describe this industrialization trajectory as 'a world of possibilities,' in the real world it is not easy to intentionally create sustainable economic gains or collective efficiencies (Schmitz, 1999) that are 'internal to the region but external to the firm' based on horizontal, forward, and backward links between small producers, non-market social coordination mechanisms, competition and cooperation, and geographical proximity.

Levy and Kuo (1991) refer to the strategy of Taiwanese firms as a bootstrapping strategy based on competition and cooperation between SMEs, as opposed to vertical subcontracting. This process presupposes social instead of technical divisions of labor and requires dense formal and informal links between independent firms, but this kind of industrialization does not proceed spontaneously through transactions between private individuals (Lazonick, 2005). For example, the Third Italy provides not only typical support such as financial and tax benefits to individual firms through producer associations and local governments, but also real services such as consulting and marketing that can be enjoyed by all but offer tailored support. Entrepreneurial and left-leaning local governments took the lead in this process. In contrast, federal government R&D support has been crucial to creating a foundation for innovation in Silicon Valley.

In theoretical terms, the world of small business associations is the world of industrial districts founded by Marshall (1890). Marshall accepted Adam Smith's concept of divisions of labor and specialization, but was very critical of innovation through the vertical technical divisions of labor in the factory system, as opposed to horizontal social divisions of labor (Marshall, 1919: 176–177).[4] He devised the conceptual framework of internal and external economies to reconcile increasing returns and competition (Marshall, 1890: 221). The latter could be enjoyed at the nation-state level through the completion of education, finance, and transportation facilities, and at the local level through the growth of interrelated industrial sectors that assist each other by being located in the same region (Marshall, 1890: 317). Marshall believed that the costs of a firm depended not only on its own production scale, but also on the development of the industry as a whole, and identified these external economies as a factor in reducing supply prices and increasing returns.[5] This contradicts the traditional classical economic view that increasing returns are due to an increase in the scale of production rather than overall economic progress.

For industrial production, specialization based on social divisions of labor is required to create this situation. In order to avoid becoming a monopoly in the face of increasing returns,[6] firms should keep pace with one another in a way that further reorganizes the social divisions of labor. This makes it possible to pool a labor force with specialized skills and form a network of specialized suppliers. The problem is that keeping pace in this way is not a trivial task, which means institutions and customs that control opportunism and collective action dilemmas are required. In this respect, the effect of knowledge spillovers is accompanied by a non-market mechanism. As Prendergast (1993) points out, in order for such a world to be economically sustainable, industrial expansion must not create economies of scale unless it brings about similar changes in the

[4] "Marx and his followers resolved to be practical, and argued that history showed a steadily hastening growth of large businesses and of mechanical administration by vast joint-stock companies: and they deduced the fatalistic conclusion that this tendency is irresistible; and must fulfill its destiny ***by making the whole State into one large joint-stock company***, in which everyone would be a shareholder" (Marshall, 1919: 176–177).

[5] Keynes (1930: 23) points out a similar situation with respect to banks: "It is evident that there is no limit to the amount of bank money which the banks can safely create *provided that they move forward in step* ... Every movement forward by an individual bank weakens it, but every such movement by one of its neighbor banks strengthens it; so that if all move forward together, no one is weakened on balance. Thus the behavior of each bank, though it cannot afford to move more than a step in advance of the others, will be governed by the average behavior of the banks as a whole—to which average, however, it is able to contribute its quota small or large."

[6] Competition would not be a necessary constraint if increasing returns from external economies were not taken into account in the economic catch-up.

way individual firms conduct business. This suggests that one path for economic growth is deepening the social divisions of labor through an alliance of small enterprises rather than deepening the technical divisions of labor undertaken by big firms, thus with the inevitability of non-market coordination being assumed.

Companies operating in such a world can be referred to as 'Marshallian firms.' Some historic examples of associations of small and medium-sized producers include the UK's industrial districts in the mid and late nineteenth century and the industrial districts of the Third Italy, a typical example of the spatial representation of 'flexible specialization' discussed as an alternative to Fordism in the 1980s. Piore and Sable (1984: 28) refer to this industrial world as a republic of independent artisans linked through dependence on one another's skills, with competition and cooperation coexisting.[7] This world capitalizes upon inter-firm organizational and technological complementarity due to the deepening social divisions of labor, which allows vertical disintegration (specialization) and horizontal competition to be socially embedded there (Lazonick, 2005). This means innovation is incremental, continuous, and based on implicit knowledge gained from experience.

In this world, problem-solving is associated with specific tools and materials. There is an apprentice system in which community knowledge and expertise build up through the experience of artisans, who apply their hard-earned skills to new problems. This learning through socialization naturally leads to cooperation and collaboration among economic entities. The problem of opportunism is less pronounced since skills are complementary and thus the problem in question tends to be solved in a self-organizing and self-reinforcing way (Herrigel & Sabel, 1999). Social divisions of labor and deepening specialization at the micro level provide the economic basis for transforming economic actors into partners rather than competitors in the production process. They may compete with one another in distribution, but opportunism and collective action are kept in check by a system of partnership, interdependence, and reputation. This

[7] As product market demand increases, the labor pool within the community becomes the basis for new entrepreneurship, and the growth of this community induces investment in regional-specific communication and distribution facilities, which in turn promotes regional concentration and vertical specialization. This makes it easy for firms to enter and creates a high level of horizontal competition (Becattini, 2002). In other words, the result is a situation that resembles perfect competition.

identity as individually autonomous but collective producers is the driving force behind building institutions to monitor collective actions such as wage cuts and the substitution of cheap materials that could undermine the community's stability and reputation, even in times of recession (see Piore & Sabel, 1984). Also, as noted by Brusco (1992) communist local governments in the Emilia-Romagna region of Italy actively implemented policies to support SMEs. Policies that offer real services such as management, marketing, and human resource training are the same as providing public goods at the local level.

Alfred Marshall believed that human nature is multifaceted, variable and can be expressed through labor, and one example that demonstrates this in reality is industrial districts (Becattini, 2002). In line with this, Marshall believed that by limiting the expansion of property or wealth to industrial accumulation, this expansion should be based on a small-scale property structure, i.e. family affection (Marshall, 1890: 228). In this respect, his views differed from Rudolf Hilferding, Max Weber, Thorstein Bunde Veblen, and John Maynard Keynes, who were critical of the accumulation of wealth based on finance as it led to unproductive consumption after the mid- to late nineteenth century or analyzed the financialized nature of capital accumulation. This suggests his worldview and its limitations which have not actively explained the world of imperfect competition dominated by large corporations.

1.2.2.2 Chandlerian and Neo-Schumpeterian Evolutionary Big Firms

Chandlerian Large Corporations

The second world of business organization can be called 'industrialization from above.' This is a world dominated by large corporations, theoretically the world of Schumpeter (1942) and empirically the world of Chandler (1977). Corporations operating in this world can be referred to as Schumpeterian-Chandlerian firms, or simply Chandlerian firms. For Schumpeter, innovation involves combining existing resources in new ways. This creates a new supply curve because it results in 'a drop – in its nature discontinuous, irregular, "unpredictable" and "historically" unique – in costs' (Schumpeter, 1928: 367). Schumpeter called this process 'creative destruction.' Accordingly, innovation implies the transformation of internal economies into external ones (Schumpeter, 1928:

384),[8] and industry-specific external economies do not drive economic development. On the other hand, Young (1928) considered externalities to be an essential component in economic progress. Private interests and social interests could encounter through a deepening division of labor, i.e. the increase in degree of roundaboutness of production.

Marshall (1890) pointed out that the envelope for successive short-run supply curves may go downward in this process of economic growth, but Schumpeter replied 'The supply curves are never theoretic curves and have not, in this sense, any theoretic meaning' (Schumpeter, 1928: 367). He argued that entrepreneurs should overcome their resistance to wide-ranging inertia or new methods. Innovation is the result of an ongoing struggle between individual entrepreneurs advocating new solutions to particular problems, and social inertia that resists change. This role of individual entrepreneurs is called 'Schumpeter Mark I.' Since then, however, Schumpeter has acknowledged the importance of cooperative entrepreneurship in large corporations (Schumpeter, 1942), which is referred to as 'Schumpeter Mark II' (Fagerberg, 2005).

Schumpeter (1954) points out that technological progress tends to be effectively consolidated through systematization and rationalization in research and management, which is achieved by a team of trained professionals who produce what is needed and operate in a predictable manner.[9] This point is in line with the context in which the multi-division (M-form) organization of American nineteenth-century large corporations emerged, as argued by Chandler (1977). Following on from Schumpeter, Chandler (1990) analyzed the role of large corporations in economic growth. Chandler showed that diversified large corporations controlled by professional managers have a competitive advantage (Chandler, 1977). He believed the major difference between American and German large corporations was their general attitude toward maintaining inter-firm relationships with their domestic competitors, with the United States demonstrating a brand of 'competitive managerial capitalism' that contrasted with Germany's 'cooperative managerial capitalism' (Chandler, 1990).

[8] Schumpeter (1909, 1928) identifies economic progress as attributable to internal economies at the micro level. The aggregation of these represents the sum of the entire economy.

[9] However, Schumpeter (1934: 20) also despised managers, seeing them as nothing more than 'supervisors.'

Chandlerian firms exist in a world where ownership and control are separate (Berle & Means, 1932), profit maximization is not the sole goal of the firm due to the existence of incomplete capital market (Jensen & Meckling, 1976), and entrepreneurial leadership by managers creates a first-mover competitive advantage. According to Chandler, investment in production, distribution, and the managerial hierarchy was crucial to maintaining the competitive advantage of large American companies.[10] It is necessary to invest in production facilities to achieve the economies of scale and scope inherent to technological development, marketing, and distribution networks. These investments allow for thee delivery of sales volume commensurate with production capacity. Additionally, they support the managerial hierarchy in coordinating production and sales, while also facilitating the planning of related investments (Chandler, 1990: 8).

Accommodating Cyert and March's (1963) argument of 'bounded rationality,' managers in this world are regarded as 'satisfiers.' Accordingly, corporate goals may focus on growth over profit maximization, and managers are expected to play an entrepreneurial role. Some of the potential downsides of this world include the concentration of economic power (Berle & Means, 1932), shirking responsibility (Alchian & Demsetz, 1972) and opportunism (Williamson, 1975), and passivity (Cyert & March, 1963), which is a result of the limitations of market discipline on firms. This limitation is partly due to the increasing number of managers resulting from the separation of ownership and control. On the other hand, the institutionalist theory (DiMaggio & Powell, 1991) explains corporate behavior in the form of isomorphism, while the neo-Schumpeterian theory (Nelson & Winter, 1982) understands corporate behavior as routine and evolution. These theories are more deterministic in understanding the harmful things found in this business world. As a result, the entrepreneurial role or discretionary power of managers is deliberated under the premise of market discipline in these discussions (Teece, 2017).

[10] There is a Chandlerian hypothesis that the competitiveness of the British economy has weakened because the United Kingdom lacked such investment due to the absence of big corporations (Chandler, 1990; Elbaum & Lazonick, 1986).

The imperfections of capital markets allow managers to play an entrepreneurial role by investing in economies of scale and scope. However, as capital markets develop, they may instead seek to maximize shareholder value, and there is a question as to whether the efficiency generated from these entrepreneurial investments is really allocative—doubts about the discretionary judgment of managers and rent-seeking. However, Lazonick (2015: 92), who interprets Chandler's argument in a progressive sense, argues that ownership and control are actually separate in large American corporations for the purpose of overcoming management rather than capital constraints on corporate growth. The decisive constraint on the growth of major conglomerates was not access to financial capital, but the management of organizational capacity to develop and utilize productive resources. Changes in the higher education system in the United States to address these issues at that time have produced a large number of professional, technical, and managerial personnel. Also, the separation of ownership and control ensures that they have a fair chance of upward mobility on the promotion ladder, and thus managerial constraints on capacity building and corporate growth have been relaxed (Lazonick, 2010). Furthermore, Lazonick and O'Sullivan (2000) argue that the goal of corporations during the neoliberal era changed from 'retain and reinvest' to 'downsize and distribute,' which has had a negative impact on investment, innovation, and jobs.

Chandler's argument assumes the stepwise view that a diversified large corporation is a more efficient form of organization than its previous steps (Whittington & Mayer, 2000: 26–27). Thus, it may miss out on non-sequential changes in the business sector. In addition, the economies of scale and scope are mostly 'internal,' while the 'external' economies of scale and scope are limited to discussions of distribution systems controlled by firms. As mentioned above, external economies can affect economic performance and economic advantage (see Marshall, 1890) but Chandler only addresses the investment problem from an internal perspective.

The Chandlerian conglomerate faces the governance problem of boundary-setting that Coase (1937) talks about. This stems from the firm-specific nature of physical and human assets according to the subdivision of tasks between organizations. When these assets are relation-specific, they have virtually no alternative value, which creates a 'hold-up' problem. The way to solve this is by integrating the system into a hierarchy (Williamson, 1975). In this sense, the efficient arrangement of tasks through vertical integration is different from the openness of industrial districts in the Marshallian world, and can actually make the business world more closed. In addition to this, conglomerates such as Korean *chaebols* are often placed under the exclusive control of a single owner-manager. This means there are costs associated with control, and as the technical division of labor deepens, monitoring or surveillance costs increase exponentially. In this regard, even if organizations expand and the technical divisions of labor deepen in order to create economies of scale, the economic benefits from these economies of scale may be offset by the costs of greater control over all complex situations. This is the limit of innovation in large corporations (see Chandler, 1977; Lazonick, 2010). If the hypothetical market world assumed by Coase (1937) becomes widespread, monitoring costs may rise due to the agency problem and the closed nature of the hierarchy of large corporations (Jensen & Meckling, 1976).

Neo-Schumpeterian Evolutionary Big Firms

Neo-Schumpeterians believe that Schumpeter's world is one of imperfect competition where the organizational capabilities of large corporations are the main source of innovation (Teece et al., 1997). We call these companies 'neo-Schumpeterian evolutionary enterprises.' According to Nelson and Winter's (1982) argument about the continuity of big firms based on organizational capabilities, innovation is cumulative in that organizational capabilities come from implicit knowledge embedded in organizational routines, collective in that firms are organizations that reproduce and represent social knowledge through coordination and learning, and involves uncertainty because firms either commit to or choose to pursue a long-term path toward capability building. As Lazonick (2005) points out, neo-Schumpeterians emphasize the importance of social and collective learning processes, but do not delve into the social conditions behind these learning processes. Lazonick (2003) views strategy, finance, and organization as the social conditions behind an innovative enterprise, and claims that assuming strategic control and financial commitment are a given, the essence of an innovative enterprise is the intra- and/or inter-firm's skill-based organizational integration that engages in collective and cumulative learning, with shop-floor workers included like Toyota.

Nelson and Winter (1982) argue that implicit expertise is a source of innovation which can be accrued within large corporations through the internal economy, without assuming the existence of an external economy consisting of a community of artisans. This assumes that large corporations could in fact consist of self-organizing communities of producers (Dosi, 1988; Sabel, 2007). According to Frederick Winslow Taylor's theory of scientific management, job or task segmentation and specialization can be achieved between management and the shop-floor within a big firm. Unlike in the Marshallian world, this specialization differs from craftsmanship because it is disconnected from a systematic understanding of tools and materials. However, detailed working standards are not clear enough to cover all situations and must be complemented by informal lived experience and tacit knowledge about processes and organizations. Nelson and Winter (1982) called this 'organizational memory' which forms a routine and is reproduced through the routine. These routines are not fixed, and can be transformed into 'dynamic capabilities' that alter according to changes in the external environment through continuous learning (Romme et al., 2010; Teece, 2017; Zollo & Winter, 2002). Dynamic capabilities can be formed in an uncertain environment, a culture of learning embedded in an organization, and an ambitious task undertaking that is suddenly decided (Zollo & Winter, 2002).

From a resource-based firm's point of view, organizational capability refers to the ability to deploy resources effectively (Penrose, 1959). This is key to creating a competitive advantage (Helfat & Peteraf, 2003). As described above, this is generally thought to consist of a combination of routines (Dosi et al., 2000; Nelson & Winter, 1982; Teece et al., 1997; Winter, 2003) and be created cumulatively and sequentially (Helfat & Peteraf, 2003; Zollo & Winter, 2002). Accordingly, routines take precedence over capabilities, and the latter cannot exist in an organization until the former is created or acquired (Nelson & Winter, 1982). For example, literature emphasizing the role of innovation in the lean production system has focused on continuous improvement activities in the workplace to eliminate waste (Womack et al., 1990). These activities are learned and coded into routines over time, which translates into organizational capabilities. Organizational capability building is therefore the process of gaining experience and articulating and codifying knowledge (Zollo & Winter, 2002). In this context, capacity building and learning processes are assumed to take place in orderly, sequential stages.

Routines are created from either extracting the organization's own direct experience or duplicating the routines of other organizations. However, there may also be other avenues of capacity building (Dosi et al., 2000; Gong et al., 2006). This understanding goes beyond the common interpretation of capability as a set of progressively and evolutionarily accumulated routines from prior experience. Gong et al. (2006) suggest two different pathways of connection between organizational capabilities and routines. The first is as described above, but the second is the opposite. This represents the formation of new capabilities through improvisational solutions to problems or opportunities, maintaining and recognizing these capabilities through iterative improvisation, and later backfilling them with supporting routines that replace them (Gong et al., 2006; Winter, 2003).

Improvisation generally refers to the temporal convergence of planning and execution with novelty (Miner et al., 2001) through agility, which is defined as quickness, lightness, and nimbleness (Highsmith, 2004). Solutions are pursued using available rather than optimal resources, within time pressure (Ciborra, 1999). Moreover, organizational improvisation as a capability goes beyond *ad hoc* activity that does not reflect a pattern behavior (Helfat & Winter, 2011). Winter (2003) clearly distinguishes between improvisation as a capability and *ad hoc* problem-solving, arguing that the latter is neither routine nor patterned while the former rests on a basis of patterned performance.

Organizational improvisation is a key foundation for capacity building and could contribute to entrepreneurial learning in organizations (Gong et al., 2006; Miner et al., 2001). This is effective in new product development (Dougherty, 1992; Miner et al., 2001; Vera & Crossan, 2005), emergencies (Weick, 1993), and conflict situations (Alinsky, 1969). Collective creativity and imagination are the keys to improvisation (Gong et al., 2006). Improvisation is also an intermediate route between *ad hoc* problem-solving and institutionalized routine formation. However, since this cannot continue indefinitely, it is necessary to translate it into routines over time. Otherwise, this could result in incompatibility with other capabilities or viable strategies (Teece, 1986, 2017), high costs (Winter, 2003), or the undermining of positive organizational performance by the repeated deployment of insufficient competencies ultimately falling into the 'competence trap' (Dougherty, 1992) which restricts the development of superior competencies over time.

Neo-Schumpeterian evolutionary big firms with organizational capabilities based on the accumulation of implicit knowledge assume the active role of managers in the same way as Chandlerian conglomerates. The role of organizational hierarchies in such firms is to extract creative contributions from employees and managers, as well as empower and deploy 'expert talent' (Teece, 2011). As mentioned above, there is an argument that market discipline should limit managers' discretionary power in order to solve the potential misallocation problem in Chandlerian big firms (Berle & Means, 1932; Jensen & Meckling, 1976). However, others believe that the capabilities and skills of managers should be utilized to build dynamic capabilities in neo-Schumpeterian firms (Teece, 2011). Accordingly the 'commonsense' corporate governance discussion that balances the two claims has widely been accepted among neo-Schumpeterians (Teece, 2017: 713).

1.2.2.3 Hybrid Inter-Firm Relationships and Post-Chandlerian Business World

Existing Inter-Firm Relationships

The final type is a hybrid of the first two business worlds. Business organizations in Germany and Japan are typical examples. Furthermore, inter-firm's various cooperation in the form of networks 'after' Chandleriean big firms characterizes this business world. This is a world in which large firms and SMEs cooperate, but there is more to it. As Thelen (2004) illustrates, the vocational training system that enabled the balanced development of large firms and SMEs in Germany is the historical product of supporting artisan handicrafts as a means of nurturing the middle class to counter organized labor in the German Empire. The result is a system of exclusive rights to skill formation among artisans, and political checks and balances between artisan labor, the political coalition of organized labor, and some factions of capital surrounding it.

In Japan, *zaibatsu* refers to a business group in which a wealthy family or its relatives maintain a closed ownership structure and operate a range of diversified businesses. This was largely dismantled by the US military government stationed in Japan after World War II, but a loosely connected system known as *keiretsu* subsequently formed. This business organization is centered around a main bank through a mutual stockholding system and served as the institutional basis for cooperation between large companies and SMEs in Japan (Aoki, 2010).

In the Marshallian world of business, it is not easy to pursue radical innovation as the organizational capabilities of individual firms are relatively weak, although Silicon Valley may be an exception. On this industrialization trajectory, it is nearly difficult for latecomers to accept or accommodate this trajectory because the demand for expansionary capital accumulation lags behind that of the Schumpeterian corporate world. As late capitalist countries in the nineteenth century, Germany, Japan, and the United States were unable to fully adopt the path of 'industrialization from below' based on industrial districts, as was the case in the United Kingdom. These nations had no choice but to compete with Britain, an advanced country by using large corporations equipped with excellent organizational and technical capabilities. However, in order for these big firms to produce large quantities of finished goods, it is necessary to procure intermediate goods such as parts and materials. At that time, mass production was impossible, and sourcing these components from other countries was difficult due to technical conditions and geo-economic circumstances. The Marshallian business world still holds true when the product differentiation is extensive and the market is too small to enjoy economies of scale like in the case of intermediate goods. A typical example is the German system of industrial organization (Herrigel, 2000).

Germany is a case where the workshops of artisans are placed under the umbrella of large corporations. Large German companies mainly produce high-quality parts and materials rather than large quantities of final goods. For example, precision machinery and fine chemistry fall under this category. Because workers are educated under formal apprenticeships organized by industry associations or vocational schools, they retain a systematic understanding of tools and materials (Thelen, 2004). Accordingly, it is possible to build tacit knowledge through experience gained at small businesses in artisan communities and undertake industry- and firm-specific learning and innovation (Herrigel, 2010). In particular, large corporations offer a lot of support akin to the public goods provided in industrial districts. This forms the basis of the German welfare system, which is based on membership. In this system, industry associations, unions, and communities play a crucial role in addressing the problem of collective action on poaching and the mobility of skilled workers (Becker, 1993). They also formally recognize the communal nature of skills to prevent a tragedy of the commons (see Busemeyer, 2009; Thelen, 2004). The German system of industrial organization involves incremental innovation and has a competitive advantage in mid-skill industries that use intermediate skills, such as automobiles and machinery (Estevez-Abe et al., 2001). However, this also entails problems of demarcation and rigidity due to the strong identity of the skilled workforce. This limits voluntary participation in team-based learning and changes in work organization (Herrigel & Sabel, 1999). In addition, employers are reluctant to bear the cost of running apprenticeships due to intensifying competition.

Unlike Germany, Japan is predominantly a system of skill and innovation that is specific to large corporations. In Japan, the degree to which skilled workers are integrated into artisanal communities and large corporations is relatively weak. This is because the intensity of 'indigenous development' was weaker than that of Germany and completely subsumed by large corporations. However, the Japanese system solves some problems about the rigidity of routines faced by large corporations through constant dialogue and communication between management and shop-floor workers, and continuously improves them. Moreover, there is a high degree of cooperation in technology transfers, including skill training between large corporations and subcontractors (Fujimoto, 1999). On the latter point, Lazonick (1990: 46–47) points out that Japanese conglomerates have achieved two levels of organizational integration through the *keiretsu* system. The first is on an inter-firm level. Organizational integration was extended to horizontally or vertically related companies through *keiretsu*. This involves an expansion of the financial control unit. The second is at the intra-firm level, where organizational integration was extended to embrace male blue-collar workers. The Japanese model has innovated and maintained a competitive advantage in technologically complex and high-fixed-cost industries such as automobiles, and this is the result of combining both vertical integration of production capabilities and horizontal cooperation to facilitate the supply of high-quality capital goods in a form of 'collective capitalism' (Lazonick, 1990: 53). This has also been conceptualized as Toyota's lean production system (Womack et al., 1990).

Workers in Japan receive extensive job training within the company in order to perform multi-functional and multi-skilled roles through job rotation and group-based improvement of problem-solving skills. However, since there is no official skill certification mechanism like the one that exists in Germany, the inter-firm mobility of employees is low, which promotes the formation of firm-specific skills (Busemeyer, 2009). Japan also lacks a strong fixed occupational identity (Herrigel & Sabel, 1999), so this does not present an obstacle to teamwork and job rotation.

Post-Chandlerian Business World

Since the 1980s, a variety of inter-firm 'network models' have taken the limelight over the rigid hierarchical management model of large corporations (Piore & Sabel, 1984). The idea that a firm's competitive advantage stems from the fluidity of its boundaries has gained acceptance (Zeitlin, 2008: 129). As a result, hierarchical management structures have declined over time. This suggests it is difficult for organizations to innovate in the hierarchical world of large corporations. As an alternative, 'pragmatist' or 'experimental' collaboration has been implemented, especially in economically developed and prosperous regions of the West (Helper et al., 2000; Herrigel, 2004, 2020; Whitford & Zeitlin, 2004).

As described above, tacit knowledge and routines are taken for granted in large organizations under the premise that self-organization or self-regulation can be undertaken within certain corporate boundaries. However, although corporate boundaries may be clearly demarcated under the premise of Coase's (1937) hypothetical world (Helper et al., 2000), in reality it is difficult for large corporations to identify the boundaries that serve as obstacles to problem-solving. In order to escape from this, it is necessary to believe that innovation comes from the outside. This systematic network formation with the outside is called 'open innovation' (Chesbrough, 2003), and amounts to innovative external corrections and amendments to intra-firm self-regulating routines. This represents the innovative capability of big firms to effectively select tacit knowledge and routines and efficient external resources. An important point here is the 'trap of routines' problem, as the building of routines involves both conflicts and incentives (Coriat & Dosi, 1998).

The enormous technological progress in the IT sector since the 1990s and the importance of the knowledge economy demand flexible and innovative organizational capabilities. This involves constantly improving existing routines through cognitive distance, similar to Japan's lean production system (Whitford & Zeitlin, 2004). Furthermore, the managerial hierarchy needs to pool information on various alternatives through 'disciplined comparisons' (Sabel & Simon, 2011: 55). In other words, large corporations can make disciplined comparisons of tacit knowledge and routines and seek continual improvement by utilizing standardization, translating implicit knowledge into explicit and formal standards, benchmarking, and simultaneous engineering. Helper et al.

(2000) call this 'pragmatic' cooperation based on 'learning through monitoring.' This type of innovation has been championed by Japanese companies such as Toyota (Whitford & Zeitlin, 2004), and minimizes agency costs by reducing the asymmetry of information between the managerial hierarchy and shop-floor workers (Aoki, 2010). It also grants greater decision-making autonomy through teamwork-based skill improvement, and streamlines inter-firm process flows through the just-in-time (JIT) system (Fujimoto, 1999).

The purpose of this is to blur the internal and external boundaries of the enterprise and challenge and improve existing routines through links between strong and weak ties (Granovetter, 1973). This pragmatic discipline implies a shift toward decentralized or empowered business organizations rather than hierarchical conglomerates (Sabel & Zeitlin, 2012). By encouraging continued reflection on existing routines, this assumes that incremental improvements over time lead to large-scale changes. Since the problem of corporate governance mainly arises from information asymmetry, pragmatist or experimental inter-firm cooperation focuses on making information more symmetrical. Furthermore, the organizational capability of a pragmatic or experimental big firm lies in integrating internal and external resources to learn and accumulate knowledge that enables continuous improvement of existing routines, and cultivate the capabilities of coordination. These practices have been illustrated as a sort of dynamic capabilities in the context of post-bureaucratic experimentalism (Herrigel, 2020).

In the 1990s, there were theoretical attempts by companies to achieve inter-firm cooperation in the form of a network as a post-Chandlerian enterprise. The aforementioned pragmatic or experimental cooperation is one of these attempts to be made. In particular, Japanese companies such as Toyota have frequently been cited as examples. However, Chandlerians take a different view of this interpretation. For example, Langlois (2002) argues that large corporations are the center of the network in Japan's business-to-business relations, and borrowing a phrase from Knight (1921: 268), this may be seen as an expansion of 'cephalization,' which means the concentration of intra-firm decision-making responsibility. Lazonick (1990) also interprets Japan's inter-firm relations as a type of horizontal cooperation based on thorough vertical integration. These claims seem to be somewhat removed from the nuances of pragmatism or experimentalism (cf. Herrigel, 2020; Sabel & Zeitlin, 2012).

One trend that has become a prominent part of inter-firm organizational structure since the 1990s is 'modularization' (Sako, 2003; Sturgeon, 2002). In this system, large suppliers leverage the advantages of relational production systems to focus on the core competencies of their customers. They do so by overcoming the rigidity of the system based on utilizing the standardization of product specifications and the codification of knowledge, allowing them to enjoy external economies of scale. This is an open inter-firm relationship of loose cooperation between customers and suppliers, where suppliers retain their own autonomy.

Langlois (2003) argues that the current situation is dominated by modularization, where coordination of decentralized production takes place through vertical specialization, general-purpose technology, and market exchanges. This is reminiscent of the Antebellum Period before the American Civil War. Sturgeon (2002) believes that the emergence of (quasi-) vertically integrated giant suppliers of standardized products is motivated by exploiting low wages. We refer to this world of networked enterprises as '(quasi-) vertically integrated, modular relational contracts.' On the other hand, Herrigel (2004) advocates a brand of pragmatic experimentation that is aimed at reintegrating conception and execution, and describes inter-firm networks with competitive advantages in high-wage regions of the West. As such, there are nuances in the interpretation of diversity in the post-Chandlerian corporate world.

1.2.2.4 Chaebol *Business Groups*

The form of business organization that served as the driving force behind Korea's catch-up in economic growth is known as the *chaebol* (Amsden, 1989). A *chaebol* is a diversified business group with no separation between control and ownership, thus combining pre-Chandlerian and Chandlerian properties. As a foundation for the growth of the Korean economy, *chaebols* achieved diversification through preferential state funding and were nurtured as 'national champions' capable of absorbing numerous technologies and increasing their market share as much as possible. In this sense, Korea's growth strategy from the 1960s to the 1997 Asian Financial Crisis was an 'investment-based' model rooted in a solid partnership between *chaebols* and the state. During that period, Korea's GDP *per capita* grew at an average annual rate of approximately 7% (Aghion et al., 2021).

This growth strategy relied on the dynamism of entrepreneurial investment by the *chaebols*, with the state providing preferential support and imposing certain requirements on firms to achieve performance targets in terms of industrial upgrading and export performance (Amsden, 1989). The state offered support through a range of policy instruments, including *de facto* preferential access to credit, manipulating the exchange rate, savings rates, suppressing wage increase, explicit and implicit bailout guarantees, and restrictions on competition from domestic and foreign direct investors (Chang, 2003; Woo-Cummings, 1999). However, the entrepreneurial dynamism and discretionary power of *chaebols* has a dual and contradictory nature. On one hand, the *chaebol* head called *chongsu* may carry out the bold investment necessary to upgrade industries and technologies, but he and/or she may also have no qualms about rent-seeking or making discretionary judgments that go well beyond rational calculations. This can lead to the *chaebols* monopolizing profits and benefits while socializing any losses under the logic of being 'too-big-to fail.'

A business group refers to a corporate business structure comprised of legally independent enterprises that engage in business activities in various industries and are bound by formal and informal ties (Khanna & Yafeh, 2007). This is a rational and efficient form of enterprise where there are imperfect market constraints on goods and services, such as in developing countries (Okazaki, 1999). In particular, when there is no proprietary technology available in related industries and potentially high returns in pre-modern industries, the diversification of business groups is highly likely to lead to opportunistic and unrelated diversification. When a business group creates superior capabilities in one sector or region through diversification, this organizational expertise can be used as a key resource in other sectors or regions. This constitutes the strategic capability of vertical integration among the subsidiaries of a business group called 'project execution capability' (Amsden & Hikino, 1994; Guillen, 2001; Lee & He, 2009). Furthermore, this concept implies that implementing more projects within the sector or region leads to an accumulation of sectoral or regional expertise, and the ensuing results can be brought within the group. For example, Lee and He (2009) use this concept to illustrate that although Samsung Electronics' entry into the Chinese market was late, the company's success can be attributed to resource sharing and coordination among subsidiaries.

The Korean government systematically filled the empty links in the domestic value chain by setting up a 'big push partnership' with *chaebols* that involved sharing investment risks in the private sector and providing support based on performance in the global market. Coordinated government investment and entrepreneurial, high-risk investment on the part of *chaebols* formed the basis for growth (Lim, 2012). Investment coordination was carried out at the national level through industrial policy, and the *chaebol* level through the strategic core competencies of 'project execution capacity' and 'vertical integration.' Korea's big push strategy was also closely linked to international trade, which reduced the coordination problems of investment, made it possible to take advantage of economies of scale, and provided learning opportunities and market testing of government policies and corporate strategies (Lim, 2003).

However, Korea's growth strategy came to an end in the face of the 1997 Asian Financial Crisis. Pressure from the International Monetary Fund (IMF) and the humiliation of the government and the people led to reform policies such as restructuring to remedy the inefficiency of *chaebols*, as well as removing barriers to entry for non-*chaebol* companies and foreign investors. Some *chaebols* were able to multiply their technological and innovative capabilities through such reforms, and non-*chaebol* companies have also followed this path (Aghion et al., 2021). The institutional changes led to more robust internal surveillance and monitoring systems, stronger external market discipline, and improved corporate business conditions. To adapt, *chaebols* adopted a brand of 'productivist neoliberalism' where symbiotic relationships are maintained with foreign financial investors and external minority shareholders. At the same time, they also faced a 'governance crisis amid excess governance' from not only the threat of hostile acquisitions by foreign capital due to the *chaebol* owner family's minority stakes in *chaebols*, but also the issue of the identity of *chaebols* as family businesses and whether management rights could be passed down to next of kin (Lee, 2013). In spite of the other changes since the financial crisis, the identity of *chaebols* as family businesses and diversified business groups has largely remained the same.

The heads of *chaebols* maintain significant corporate control over the residual claims they hold, mainly because of the absence of a clear separation between ownership and management, as well as the pursuit of non-related diversification through the roundabout cross-shareholding arrangements. However, there is a so-called 'gap' between ownership and control (Kim, 2003). This means that even if the head only holds a small stake in affiliates, they can exercise significant control through the stakes of related persons such as family members and cross-shareholding between affiliates. *Chaebols* formed through the pursuit of unrelated diversification can also utilize quasi-internal capital markets. This means that the headquarters seek the efficient use of resources by owning and controlling each division as a capital investor. However, if there is sufficient idle cash flow within the *chaebol* that can be used by the head, bureaucratic investment decisions can lead to overinvestment. It is virtually impossible to monitor the quasi-internal capital market from the outside, so *chaebols* can effectively neutralize external surveillance and monitoring systems such as banks by managing affiliates through internal financial transactions (Song, 2007).

In the framework of real options, the Anglo-American model of shareholder capitalism can lead to 'Type 1' errors that fall short of the average market rate of return due to short-term-oriented investments and dividends stemming from serious agency problems between shareholders and managers. However, in the German and Japanese stakeholder model, where insider control is made possible, long-term-oriented investment can capture additional option value based on the contingent and complementary relationship between managers and employees. This option value is supposed to pursue a certain degree of rationality through an internal or external monitoring system, but investment decisions can become decidedly irrational when the line is crossed. The *chaebol* system has still weak external checks, which can lead to companies committing 'Type 2' errors if they believe they are too-big-to-fail. The overinvestment of *chaebols* that caused the 1997 Asian Financial Crisis is one such example. In this way, the entrepreneurial dynamism of *chaebols* is a double-edged sword for the Korean economy.

The *chaebol* model is accompanied by a hostile management system, relying on controls rather than economic incentives. Prior to Korea's democratization in 1987, if the cost of labor controls is transferred outside the company through the market discipline like the threat effect of dismissal resulting from relative abundance of labor in a surplus population, or the state's oppressive control against labor, the *chaebol*'s cost may not be large. The *de facto* authoritarian state bore the *chaebol*'s cost of labor controls (Song, 1999). However, this oppressive system was threatened after the 'Great Labor Struggle of 1987,' and the state was relegated to the role of an indirect mediator. Since the mid-1980s, the *chaebol*'s cost of labor controls has rapidly increased due to weaker market discipline arising from labor shortages and the state's less interventionist stance. In the face of this, *chaebols* have used both carrots and sticks, providing workers with comparatively high wages and improving corporate welfare.

As a result, Korea has seen widening wage gaps between firms of different sizes, leading to a more segmented and dual structure in the labor market. For example, *chaebol* conglomerates have attempted to include workers at large corporations as members of insider control. In addition, as the Korean labor market has become more flexible since the Asian Financial Crisis, the need for firm-specific skill formation has declined and it is now easier to replace production workers through numerical flexibility. This means companies have greater incentives to implement highly flexible, automated production systems that separate concept from execution by combining the latest automation technology from engineers with semi-skilled or unskilled blue-collar workers.

As mentioned above, Amsden's (1989) analysis of the Korean economy's catch-up before the 1980s emphasized the importance of managers, especially engineers who exercised considerable decision-making power in production technology and R&D to absorb foreign technologies. Lazonick (1990) reached a similar conclusion after analyzing managerial capitalism in the United States. The technical manpower training policy, including engineers, has been a subordinate of economic and industrial policy, and maintained high-tech-oriented, supplier-centered framework in Korea. The Korean government has implemented policies to train technical professionals in order to grow and support the light industry in the 1960s, heavy chemical industry in the 1970s, high-tech industries such as semiconductors in the 1980s, strategic industries such as mobile phones and automobiles in the 1990s, and cross-fertilizing industries and key enabling technologies such as information and business technology after the 2000s.

Economic and job training plans were carried out in tandem, with the latter focusing on a preemptive oversupply strategy in response to demand (Jeong, 2021). It is well-acknowledged that the excellent technical professionals supplied in this way have contributed greatly to the rapid learning and efficient use of technology imported from developed countries in Korea, as well as subsequent improvements (Amsden, 2001). The increased number of scientists and engineers created a highly general-purpose knowledge base that could adapt to different situations (Di Maio, 2008). In order to improve a country's technological capabilities, it is necessary to invest in higher education and universities that produce excellent scientists and engineers.

However, importing advanced foreign technology is never an easy process, and the import of technology is not enough to catch up with advanced players in the absence of investment in education, training, and R&D (Dosi et al., 1995). Amsden (2001) argues that Korea's technological catch-up was possible because of the integration between industrial policies and policies to train technical professionals. However, the rate of students entering tertiary education in Korea already reached the world's highest level in the 2000s, with a high proportion of STEM majors. This means that although there is a sufficient quantitative supply of technical labor, there is always a shortage of up-to-date technical professionals due to the emergence of new knowledge-based technologies. As suggested by Amdsen (2001), the success story of Korea continues to obsess over new cutting-edge technologies and maintain the unilinear view of supplying excellent technical professionals to support them.

When comparing R&D with workplace manufacturing capabilities, it was the internal accumulation of engineering capabilities in production design and process technology that played an important role in Korea's successful catch-up (Amsden, 1989). As noted by Cho (2014), there has been very little room for functional and hierarchical integration of various skills ranging from R&D to the workplace, and the integration has been largely led by engineers and managers who exclude the shop-floor, as is the case in the United States. Kim et al. (2011) observe that process engineers have effectively made up for the skill deficiencies in shop-floor labor. As a result, Korea has seen disproportionate development between shop-floor skills and technical capabilities in product technology, production design, and process technology. Unlike in Japan, this sharp divide between technology and skill makes organizational innovation difficult (Hattori, 1986).

As argued by Amden (1989) and Hattori (1986), Korea's technological advancement has mainly been achieved through continual exchanges of information and knowledge between engineers and technicians rather than engineers and shop-floor workers, as is the case in Japan. According to Hattori (1986), Korea's success in absorbing advanced foreign technologies before the 1997 Asian Financial Crisis was due to the relative ease of learning standardized and mature technologies, efforts to absorb such technologies on the individual level, and the long tenure of most employees. The OECD (2019) also presented a study of the Korean production system that found a lack of cognitive and non-routine skills among shop-floor laborers and that companies prefer to employ engineers.

Chaebol conglomerates reduce costs and accumulate technology internally through quasi-vertical integration of core materials and parts from major affiliates. New affiliates tend to grow rapidly in the short term through such insider trading, and can be used to solidify the *chongsu* family's succession structure of weak corporate governance or even engage in private defrauding. Korea's *chaebols* have a closed structure with a clear divide between the inside and outside of the organization. This entails considerable transaction costs in organizing the social divisions of labor, so when investment in specific assets is required, *chaebols* typically prefer mobilizing internal resources or bringing external resources into the system over exploring or exploiting new or existing relationships with external companies. This is different from the vertical integration of large European or American companies in that quasi-vertical integration takes place in a group of companies rather than within a single enterprise. It can also be distinguished from Japan's *keiretsu* in terms of quick decision-making centered around the head of the group.

By taking advantage of their monopsony on demand, *chaebols* are able to extend quasi-vertical integration beyond direct control over internal affiliates to cooperative external suppliers. This grants *de facto* direct control over the corporate business ecosystem, irrespective of legal boundaries. We call this system 'extended quasi-vertical integration.' This implies that *chaebol* conglomerates can exercise complete control over the external business ecosystem through a vertical and hierarchical transaction structure. *Chaebols* thus implement modularization in parts sourcing in a closed manner by selectively stratifying and excluding external suppliers with help from the system they created.

Extended quasi-vertical integration provides a way to both transfer and share risks with external vendors. For example, if a *chaebol* experiences increased uncertainty in market demand due to rising raw material prices, a fragmentation in demand, or a general economic crisis, they are able to pass this risk on to external suppliers through multiple orders from core *chaebol* affiliates and periodic unit price cuts. At the same time, *chaebols* also provide a stable market for external partners because they purchase parts and materials in a long-term relationship, offer staffing, expertise and investment support, and sometimes absorb market fluctuations to a certain extent. This helps external partners to stay afloat in times of uncertainty. However, since this is still a closed system at heart, it can distort market competition by reducing fair trade or market opportunities for new entrants and also cause problems with the concentration of economic power (Jeong, 2022).

1.2.3 Automotive Production Systems

1.2.3.1 Prior Literature on Automobile Production Systems

The mainstream of research on automobile production systems over the past 30 years has been aimed at identifying the main characteristics of the lean production system and examining issues involved in introducing and transplanting such a system, as well as its variants. One example that captures this discussion is *The Cambridge International Handbook of Lean Production: Diverging Theories and New Industries around the World*, edited by Janoski and Lepadatu (2021). Womack et al. (1990) argued that Toyota's lean production system[11] was the new 'single best way' and could be successfully transplanted to new environments. Representing the Groupe d' Etudes et de Recherches Permanent sur l' Industrie et les Salariés de l'Automobile (GERPISA), a European-focused network of automotive researchers, Freyssenet et al. (1998) and Boyer and Freyssenet (2002) critically responded to these MIT-International Motor Vehicle Program (MIT-IMVP) claims. They disagreed with the 'single best way' argument and showed the strategic differences and diversity in the global automobile industry through a new framework of production systems.

On the other hand, Dore (1992) emphasized the embedded sociocultural nature of the Japanese automotive production system and argued that this constrained its portability. The Toyota system stands out for its links between the production process, work organization, and social organization system (Jürgens, 1991, 1992). This implies that the production system maintains a certain degree of self-regulation and institutional complementarity with the closed social integration and control systems in Japan. There is a difference in the degree of 'application,' which represents the complete functional and spatial transfer of a production system, and 'adaptation,' where the system is changed in line with specific local circumstances. Accordingly, Abo (1994) argues that a hybrid form of lean production has emerged. Hampson (1999) claims that Toyota's lean production system is a compromise between the two opposing goals

[11] In this book, the lean production system and the Toyota production system are used virtually synonymously. Janoski and Lepadatu (2021) point out that this is not the case and discuss different variations of lean production systems.

of pursuing balanced production through *kaizen* and eliminating waste arising from humans and machines. He also notes that this is influenced by the ability of labor–management relations to resist practices such as forced overtime on short notice and 'management under stress' (Babson, 1995). In fact, in the process of transplanting Japan's lean production system to the United States, it appears that production workers were exploited through just such practices (Babson, 1995) or 'chaos on the shop-floor' (Juravich, 1985). The emphasis on forced overtime, work intensification, outsourcing, and the use of temporary workers is one negative aspect of the lean production system (Lepadatu & Janoski, 2018).

Although the lean production system has both positive and negative properties (Janoski & Lepadatu, 2021), it is clear that the Toyota system is currently the dominant model in the global automobile industry. In this sense, in order to understand Hyundai Motor's own production system as the main focus in this book, it is helpful to draw comparisons with Toyota.

In terms of Korean literature on the automobile production system, the most important issue of discussion in the 1990s was whether or not to 'Japan-ize' the Korean automobile production system. For example, Jung (1995) and Joo (2007) believed that many aspects of the Japanese system had been incorporated into Korea's production system. On the other hand, Lee (1994) and Jo (1992) emphasized the Fordist nature of the domestic production system.

Since the 2000s, studies have begun to view the production system of Hyundai Motor, which represents the Korean automobile industry, as having its own properties. Cho et al. (2008) described the Korean labor and employment regime as a mixed market economy and positioned Hyundai Motor's production system in that context. Kim (2010) examined modular production of parts and components as a key element of the Korean automobile industry since the 2000s. Kim's study came from the perspective of the value chain, and explained how modularization led to polarization in the employment system. Hyundai Motor's modular production is different from the European and American systems, which feature a strong arm's length market relationship, and the Japanese model's central focus on design, in that it is promoted mainly by the affiliate company Hyundai Mobis. Jo (2016) pointed out that it would be impossible for Hyundai Motor to adopt the Toyota production system, which requires skill improvement and active participation from shop-floor workers, due to hostile labor relations at Hyundai Motor (Lee & Jo, 2007).

1.2.3.2 Concepts and Types of Automobile Production Systems

There is no widely accepted definition of a production system. In a dictionary sense, it refers to any method for creating goods or services using resources. In other words, workers, technology (knowledge), and tools (machines or facilities) interact to transform materials into something of value. If we accept this definition, the elements that make up the production system are man, machinery, materials, and technology (knowledge). Different production systems emerge depending on how these elements are combined. Skinner (1985) distinguishes between tools (hardware elements) and infrastructure (organizational elements) as components of a production system. Jürgens (1995: 298) refers to a production system as 'a set of work and process organization forms.' Industrial sociologists view this relationship from the perspective of control and exploitation, while industrial engineers seek to optimize this relationship to improve productivity.

There have also been attempts to define production systems as the links between production control and work systems and social organizations (Dohse et al., 1985), and as structural frameworks for production organizations and social processes that encompass corporate strategy (Jürgens, 2002). This brings social organization, corporate strategy, and governance into the discussion. One such attempt is the very comprehensive production system or regime concept of GERPISA (Boyer & Freyssenet, 2002). This includes corporate governance, profit strategy, and macroeconomic structures surrounding growth and distribution within the concept of a production system. The GERPISA group leans on this concept to show the emergence of hybrid forms of production systems (Boyer & Freyssenet, 1995) and strategic differences between major automakers in the global automotive industry (Boyer & Freyssenet, 2002). As mentioned above, this view is also quite critical of the theory that views the Toyota production system as the 'single best way' (Freyssenet et al., 1998).

Automobile production systems can be broadly divided into three categories (Clarke, 2005). The first is a mass production system that combines Taylorism and Fordism. The former reorganized the division of labor by fragmenting artisan skills through the introduction of scientific work principles based on time and motion studies on work processes. It defines the relationships between operators and works on the target of work chains and tasks in the production process. The result is a highly fragmented division of labor and a separation between concepts and execution (Braverman, 1974). Artisan skills are devalued through the use of scientific methods to determine task content, resulting in highly repetitive standardized tasks. Standards defining tasks are created externally by engineers, and workplace expertise is not integrated into everyday jobs (Dohse et al., 1985). In other words, there is a clear separation between the roles of the shop-floor and engineers. Standardization increases the level of control over jobs and workers. In this work structure, hierarchical concentration and control of skills replace horizontal empowerment.

Fordism is a system of mass production that combines Taylorist scientific management with technological progress (Braverman, 1974). Mass production was made possible by initially relying on one standardized low-cost product and using standardized interchangeable parts. Moving assembly lines were introduced to implement the principle of process flow. This combines Taylorism's work standards with Fordism's standardized production flow, or process standards. As such, the components of this mass production system are standardized work, interchangeable parts and components, and a moving assembly line. Fordism focused on mechanization or automation, where parts and labor become replaceable through the integration of Taylorism.

Fordism was also able to exploit economies of scale through social control. High-wage policies that selectively target employees were used as a pecuniary reward for workers who endured difficult and monotonous working conditions (Braverman, 1974: 103). Ford used Taylor's scientific management method to control the sequence and timing of work, which went beyond labor controls and even penetrated into the private sphere of workers (Clarke, 2005: 88). In this way, the system of mass production linked process, work, and social organization. As Dohse et al. (1985) point out, the worker thus became an 'appendage of the production apparatus,' which involved oppressive controls over the labor process. This system is generally characterized by a low level of responsibility for shop-floor labor, engineers in charge of process innovation, and human resource management based on a hire-and-fire strategy in which workers are fired without their consent and re-hired under new conditions. Skills are acquired through on-the-job training, and industrial relations are marked by distrust and hostility (Clarke, 2005; Jürgens, 1995, 1998).

The second type of system is Volvo's Uddevalla production system, which was designed to overcome the labor alienation present in Taylorism. In the 1970s, attempts to transform mass production systems that alienated shop-floor workers into systems that match economic efficiency and social conditions emerged as a 'humanization of labor.' This led to Volvo's Uddevalla production system in the 1990s. This model deliberately rejects the use of standards to define work. In other words, it gives individual workers and teams the right to autonomously organize their own work. Each team determines their own pace of work, and jobs are based on general tasks and long work cycles (Sandberg, 1994). However, this production system was overly idealistic and did not last for very long.

The final model is the Toyota production system, popularized by Womack et al. (1990) as a lean production system. Since the 1990s, the lean production system has become the main point of reference for production systems in the global automobile industry. However, Kenney and Florida (1988: 123) strongly criticized Toyota's plants in the United States as 'hyper-Fordism, a more advanced and exploitative version of Fordism.' Adler and Cole (1993) used contradictory words for New United Motor Manufacturing Inc. (NUMMI), a joint venture between GM and Toyota, describing it as both a 'learning bureaucracy' and 'democratic Taylorism' in which learning takes place in hierarchical organizations with a clear division of roles between shop-floor workers and engineers.

The Toyota production system stands in the tradition of Taylorism and Fordism, but is not a simple remake of these models. For example, standards and capabilities are created inside the workplace and evolve dynamically through continuous improvement. Work is team-oriented, but based on highly repetitive tasks and short work cycles. As implied by the word 'just-in-time,' the pace of work is regulated through relationships (Fujimoto, 1999; Jürgens, 1994; Ohno, 1988; Shimizu, 1995). This shows that social organization and above all work organization correspond to the process chain (Dohse et al., 1985). If the principle of the Toyota production system is to reduce costs by fully eliminating waste, constant pressure to cut costs can lead to 'management by stress' (Babson, 1995) or forced overtime (Hampson, 1999).

The Toyota production system is different from a mass production system. While industrial engineers set work standards in traditional mass production systems, continual improvement of work standards and capabilities is the primary responsibility of shop-floor workers at Toyota. A link between performance incentives and efficiency improvements has also been established through continued improvements to operations, which is a major factor in Toyota's success (Jürgens, 1994). If external pressures such as *kanban*, just-in-time controls over production and inventory levels, i.e. constraints on the overall process flow are not smoothed away, the wages of shop-floor workers may potentially decrease, so the workers have no choice but to cooperate. Toyota's production system also promotes multifunctional or multi-skilled team-based works in connection with a system of compensation for corporate goals and makes extensive use of the intellectual capacity of workers.

In the mass production system, supplier management relies on arm's length market transactions, while in the Toyota model, a *keiretsu*-based parts supply network is integrated into the production and product development process. This served to smooth out relations between Toyota and its suppliers (Kenney & Florida, 1988). The Toyota production system also consists of a set of firm-specific routine capabilities (Fujimoto, 1999) that have evolved over time through learning and are cumulative in nature (Nonaka, 1990). Although social controls over work organization were established even in the mass production system, this was further strengthened in the Toyota system (Clarke, 2005; Dohse et al., 1985). In this respect, it can be viewed as an evolution of the Fordism principle rather than an alternative to the mass production system (Jürgens, 1994), with working within a complex system of the worker's self-regulation, its active participation in the labor process, and social integration and control.

1.2.3.3 Operationalization of Automotive Production Systems

The next step is to operationally define the concept of production systems to be used in this book. Womack et al.'s (1990) research based on the MIT-IMVP group's working hypothesis tends to overlook the macroeconomic and social context and tends to focus on technology-oriented thinking confined to micro-level work organizations. On the other hand, the concept used by Boyer and Freyssenet (2002), which represents the GERPISA group, is marked by a core profit strategy that includes scale, scale and diversity, innovation and flexibility, quality, constant price reductions, and diversity and flexibility. This profit strategy interacts with goods, labor, and financial markets, and is influenced by the growth and distribution patterns of the national economy and the international division of labor. This in turn is related to corporate governance in a broad sense that encompasses each firm's product strategy (markets and types of products sold), work organization (skills and supplier relationships), financial strategy, and employment relationships (roles and worker compensation). In other words, a production system is a compromise between major actors at the micro level (shareholders, managers, unions, and suppliers), and a combination between the broad governance structure and growth regime at the macro level. This combination takes on different properties from country to country and company to company. This definition serves to deepen our understanding of the links between corporate strategy, markets, and employment relationships.

On the other hand, automotive production systems are defined by two aspects (Jürgens & Krzywdzinski, 2016). One is the overall goal resulting from corporate strategy and the market environment, and the other is the social process. This means a production system is a structural framework for organizations and social processes. Herrigel (2020) defines this as a system that embodies goal setting among stakeholders to continuously achieve innovation, cost reductions, and learning.

There are a number of analytical frameworks that link the micro-firm level with the macro national economy level. A leading example of this is Aoki's (2001) comparative institutional theory for A, J, and G type corporations. This model seeks to classify corporate governance structure according to differences in the sharing of information and the learning of knowledge at the corporate level, combined with the characteristics of each country. The discussion of varieties of capitalism also begins with an analysis at the firm level and the attributes of each firm are linked to the characteristics of each country. The type of capitalism, roughly divided into the so-called free market economy and the coordinated market economy, was inferred and reduced from the analysis at the firm level (Hall & Soskice, 2001). Based on the economics of convention, Storper and Salais' (1997) worlds of production theory also links mid-range industry-level practices with micro-firm level analysis. Capitalizing upon the nature of product and market, they come up with four types of industrial world and then raise several hypotheses. However, these studies have limitations in that they understand the relationship between the macro- or mid-level and micro-level in terms of structural isomorphism. The prevailing view is that the relationship between the micro and macro levels corresponds to or is connected in a one-to-one relationship.

On the other hand, Boyer and Freyssenet's (2002) discussion of production systems notes that where the micro-condition of governance compromise and the macro-condition of growth regime consistently match, different types of production systems or regimes can exist because the latter is a necessary but not sufficient condition. Accordingly, the one-to-one relationship between the micro and macro levels cannot be maintained. We think this argument is a useful analytical framework in that it considers a broader range of diversity. However, this does lack a detailed analysis of financialization, globalization, and the political economy of national growth regimes and international divisions of labor (see Freyssenet, 2005; Lung, 2005).

The Hyundai Motor production system, the focus of our analysis, is inseparable from the *chaebol* governance structure. This structure exerts great influence not only at the micro level of technology and work organization, but also at the macro-institutional level. To incorporate this property and place the concept of production systems in a wider socioeconomic context, we redefine the Boyer and Freyssenet's (2002) and Jürgens and Krzywdzinski's (2016) definitions of production systems at the mid-range level. As pointed out by Lung (2005), it is true that the growth regime at the national level and the external international environment have a significant impact on production systems.

In this book, however, we place more emphasis on factors that actually affect the production system such as *chaebol* corporate governance, product strategy, work and process organizations, and employment relations. That is, the production system is defined as a set of a firm's product strategy, work and process organizations, and employment relationship under a particular governance structure. In addition, keeping in mind the social and economic effects of governance, it seems necessary to delve into the connection between work organization and social processes emphasized by Jürgens and Krzywdzinski (2016). Since corporate governance structure in this context refers to a framework that regulates the allocation of roles and distribution of profits among various stakeholders (Aoki, 2006), we believe we can shed light on the socioeconomic meaning of production systems by examining the relationships between labor and management, shareholders and owners, and main contractors and subcontractors.

1.3 Conceptualizing Hyundai Motor's Agile Production System as Authoritarian Experimentalism

1.3.1 Conceptual Map of This Book

So far, we have extensively reviewed discussions on Korea's catch-up, forms of business organization including *chaebol* governance, and automobile production systems to explain the two sides of Hyundai Motor's remarkable growth since the end of the 1990s (see Fig. 1.1). The arguments of Veblen (1915a) and Gerschenkron (1962) are particularly important in relation to the concept of economic catch-up. The former emphasizes the 'merits of borrowing,' an open benchmarking strategy that may ensure mechanistic rationality or conception by comparatively examining the catch-up processes of Britain, Germany, the United States, and Japan (Veblen, 1915a: 21; 1915b). This allows conventional habits and ways of thinking to free themselves from impediments to the production of goods and services by encouraging greater liberty and mechanistic rationality. For example, Veblen criticizes the militaristic paths of the dynastic states of Germany and Japan, the difficulties of capital accumulation caused by conspicuous consumption of the British elite, and the financial nature of American capitalism. In this way, his institutionalist discussion of a comparative catch-up process implies a vision of developmental transformation that avoids such pitfalls.

On the other hand, the discussion of Gerschenkron (1962) is disconnected from stepwise, successive, and sequential ideas. Gerschenkron emphasizes the importance of non-market mechanisms in expanding and building up the internal capabilities to catch up, arguing that 'leapfrogging' is required to successfully catch up to advanced players. Although his argument has weak political and economic implications, it exemplifies the possibility of latecomers being able to catch up.

When discussing the catch-up process, we strive to avoid a sort of dichotomy between market and non-market, continuity and discontinuity, inside and outside, and so on. The discussions of Veblen (1915a, 1915b) and Gerschenkron (1962) are not contradictory. We accept that internal capability-building is a non-market, trial and error process of perseverance, while at the same time fully acknowledging the advantages of benchmarking through the transfer of external skills and market knowledge in an age of globalization and technological progress.

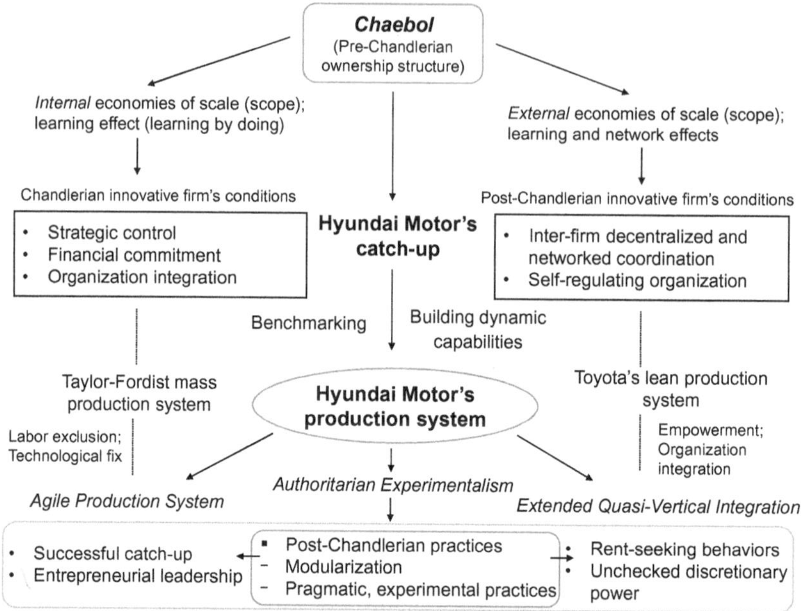

Fig. 1.1 Conceptual map of this book (*Source* Author's own creation)

In the preceding section we discussed Marshallian SMEs, Chandlerian conglomerates and neo-Schumpeterian evolutionary big firms, and existing hybrid forms of vertical and horizontal cooperation between the two and post-Chandlerian business world. Lazonick (2005) suggests strategic control, financial commitment or patient capital, and organizational integration as the conditions that lead to innovative firms from the Chandlerian perspective. However, his discussion somewhat undervalues Marshallian business practices such as horizontal cooperation between large corporations and SMEs. On the other hand, Sabel (1994, 2006) lists decentralized, networked inter-firm coordination and self-regulating organizations as the properties of pragmatic, experimental inter-firm cooperation, and views post-Chandlerian corporate practices as best practices. Sabel believes the current industrial world consists of a set of horizontally cooperative links between large enterprises and SMEs, and thus highlights the significance of the corporate business ecosystem. In contrast, Sturgeon (2002) noted the emergence of large modular suppliers that leverage external economies of scale through vertical disintegration in a post-Chandlerian context. This demonstrates the overwhelming market power of Chandlerian conglomerates in the process of configuring the global value chain and is a *de facto* extension of the Chandlerian corporate business world.

Schumpeter (1928) highlights the sources of internal economies without considering external economies as a source of innovation, and Chandler (1977, 1990) does the same. However, the emergence of decentralized, networked firms based on various vertical and horizontal links between large and small firms since the 1980s emphasizes the importance of external economies of scale (scope) and geographical agglomerations or clusters as a source of growth and a form of collective efficiency (Schmitz, 1999). It is this point that is connected to the business ecosystem and reflects the power dynamics between businesses.

Accordingly, post-Chandlerian practices such as modularization and the expansion of horizontally cooperative inter-firm pragmatic capabilities as part of the building of dynamic internal capabilities presupposes a corporate business ecosystem. This ecosystem is a key component of automobile production systems. Toyota's production system links production process flows with demand fluctuations not only within the company but also between the firm and its suppliers across the enterprise ecosystem, and this is based on dynamic capability-building through continuous improvement in shop-floor skills (Fujimoto, 1999). As a result, this implicitly assumes a neo-Schumpeterian evolutionary big firm.

As a latecomer, Hyundai Motor has succeeded in catching up to more advanced competitors by drawing on properties of both the traditional mass production system and Toyota's production system to create a model of its own. However, catching up is only possible by jumping or skipping stages in the traditional path or creating a new route entirely. Hyundai Motor benchmarked the latest technology practices at the time to promote an automation-centered technocratic solution that excludes shop-floor labor. In Hyundai Motor, engineers as a core resource of Chandlerian conglomerates were, though not entirely, empowered to create dynamic organizational capabilities in a way different from skill formation of shop-floor workers as in Toyota. The company has also acquired economic benefits from modularization in a way different from existing automobile production systems, and it successfully caught up by extending quasi-vertical integration applied to affiliates across the entire corporate business ecosystem. This has been achieved by leveraging its monopsony on demand from the position of a *chaebol*. However, this dynamic process could become a double-edged sword due to the potential rent-seeking behavior inherent in the *chaebol* system and the bold and discretionary decision-making of unchecked *chaebol* leaders turning into ruins in hard times. We call the automobile production system that emerges from this dynamic but contradictory process an 'agile production system,' and the main organizational principle 'authoritarian experimentalism,' as the key conceptual blocks and constructs used in this book are mapped in Fig. 1.1. A detailed discussion of these concepts follows in the next section.

1.3.2 Hyundai Motor's Agile Production System: An Authoritarian Experimentalism

This book attempts to reveal how Hyundai Motor was able to catch up advanced overseas competitors since the late 1990s, looking at both the success story and the negative aspects that lie behind it. The three key concepts in this process of authoritarian experimentalism are catching up, *chaebol* governance, and automobile production systems. How can this conceptual framework be used to understand Hyundai Motor's success?

1.3.2.1 Agile Production Systems

The historical origins of the agile production system can be found in the US automobile industry in the early 1990s when some top executives and analysts in the United States devised to counter Japan's lean production system. Funded by the US government, the Iacocca Institute ran an 'agility forum' to devise manufacturing strategies that would lead the twenty-first century (Sharp et al., 1999). This forum sought to obtain a new competitive advantage for US manufacturers by building an agile production system to cope with the uncertain environment (Kidd, 1994).

Table 1.1 compares the characteristics of the agile production system with the mass production and lean production systems. Above all, US automakers attempted to overcome their competitive disadvantage against Japanese automobile companies through bold investment in equipment and information technology and entrepreneurial leadership from the CEO. This model assumed an 'ideal' production system led by an autonomous work team comprised of skilled professionals. However, the agile production system outlined in this table was nothing more than an unobtainable 'dream' for US automakers, but a 'dream' suitable for

Silicon Valley startups, and ultimately failed. The filing of bankruptcy protection by GM and Chrysler during the Global Financial Crisis in 2008 signified the failure of the American version of the agile production system.

The lean production system sought to eliminate waste through efficient management in a relatively stable environment and reduce costs and increase added value through the productive use of limited resources to cope with fierce competition. In contrast, the agile production system focuses on improving the firm's ability to respond quickly and holistically in a rapidly changing environment and increase overall added value by seizing advantageous opportunities. Although the two production systems share many specific elements, the agile production system has more room for managerial discretion and flexibility in allocating resources and formulating strategies to cope with uncertainty.

Table 1.1 Comparison of Iacocca Institute Production Systems

	Mass production system	*Lean production system*	*Agile production system*
Driving forces	Economies of scale, stable markets	Elimination of waste, predictable markets	Economies of diversity, uncertain markets
Products	Standardized products	Products with a variety of specifications	Fragmented products
Product development time	Several years	Several months	Several weeks
Human capital	Division of labor, subdivided hierarchy	Human capital investment, team organization	Autonomous working teams, overcoming business obstacles
Organization	Hierarchical division of labor	Horizontal team system	Empowerment
Investment	Equipment and facilities investment	Technology investment, reducing waste	Information and staff investment, high-quality equipment, and infrastructure
Supplier relationships	Arm's length market relationships	Long-term cooperation	Short-term cooperation

Source Author's own adaptation from Sharp et al. (1999)

For example, quality management is one of the key features of Hyundai Motor's agile production system, as will be described in Chapters 2 and 3. Based on the view that quality improvement was essential to respond to customer demands, Chairman Chung Mong-Koo focused Hyundai Motor's firm-wide capabilities on quality management. Unlike the lean production system, the agile production system emphasizes the ability to quickly reorganize organizational capabilities with the help of the CEO's entrepreneurial and driving leadership. To stabilize the quality of automobiles before mass production, Chung chose to build a pilot center despite the considerable cost. The pilot center served as a physical space and institutional device that enabled agile responding by having engineers learn through improvisation, as will be discussed in Chapter 3. Since it was possible to quickly redeploy resources, the pilot center became a space where different stakeholders experimentally, intensively, and collectively solve problems that arose during new product development.

Agile production systems focus on appropriating and exploiting advantageous opportunities as an overall response and strategy to a highly volatile external environment, allowing for management to make prompt decisions and exercise discretion. *Chaebols* have the entrepreneurial discretionary power to create and appropriate real option value in response to situational and contingent changes through concentrated rapid and bold decision-making. However, the sunk cost nature of such decision-making can lead to the possibility of committing a Type 2 error, where companies believe they are 'too-big-to-fail' or the resulting losses are socialized. *Chaebols* operate a variety of teams to improve product quality, make quick and agile decisions by maximizing available resources, and leverage advantages as conglomerates by mobilizing Chandlerian resources on the basis of pre-Chandlerian ownership structure. For example, Hyundai Motor extends quasi-vertical integration applied to affiliates throughout the supply ecosystem in order to implement modularization, utilizing its monopsony on demand in the parts market, as will be described in Chapter 5. On the other hand, although limited in scope, there have been pragmatic experiments such as collaboration and deliberation between various stakeholders by delegating certain authority to engineers. The bottom line is that, paradoxically, post-Chandlerian practices such as modularization and pragmatic collaboration were agilely accommodated through the *chaebol* system.

Compared to the lean production system, the agile production system favors technocratic solutions. Hyundai Motor is a typical example of this, as is Tesla. Given the social and institutional limits of the shop-floor due to confrontational labor–management relations, Hyundai Motor has focused on engineer-led technocratic solutions instead, as will be detailed in Chapter 4. This means the company has failed to achieve the 'reintegration' of concept and execution envisaged by the pragmatic experimental enterprise. In a different sense, Hyundai Motor has a strong pragmatic mindset that makes it easy to benchmark new technological opportunities when setting up a new production system, as engineers do not have a strong occupational identity and shop-floor workers are not as skilled as in other competitors. In other words, Hyundai Motor actively utilizes the 'window of opportunity' through automation and informatization along with the introduction of the latest technology and its benchmarking.

Catching up was easier thanks to standardization, modularization, and the falling prices of capital goods. The trend toward modularization is expected to accelerate further in the era of electric vehicles, making Veblen's (1915a) argument on machine-technology more relevant. However, modularization is not entirely open to all parts suppliers due to the integrated architecture in the automotive industry. Instead, it is carried out between a specific lead firm and a parts supplier in a closed fashion. This means engineers need to be able to coordinate and design the major roles and functions within this architecture. Therefore, the successful developmental catch-up process still needs both internal capability-building and nimble benchmarking in the automotive industry, and both are not mutually exclusive. In this respect, the problématiques of Veblen (1915a) and Gerschenkron (1962) are not mutually contradictory.

We think that the problématique of US automakers, who found the key to competitive advantage in agility, is still effective in explaining the twenty-first-century paradigm shift in the automobile industry. In the twenty-first century, Hyundai Motor stood out as a finished carmaker by implementing elements of the agile production system that failed in America. Instead of promoting skill formation in shop-floor labor, Tesla has also favored engineer-led technocratic solutions and giving entrepreneurial discretion to the CEO and managers. While Toyota Motor has pioneered the lean production system, it could be said that Hyundai Motor has led the implementation of the agile production system.

1.3.2.2 Authoritarian Experimentalism

During Hyundai Motor's second catch-up period after the late 1990s, the company was a contradictory but innovative firm that embraced post-Chandlerian practices (such as pragmatic experimentation and modularization) by utilizing Chandlerian resources (the corporate hierarchy and engineer-managers) on a pre-Chandlerian basis of ownership (non-separation between ownership and management). In this section, we focus on how to describe this process in a word. Adler and Cole (1993) defined the attributes of the NUMMI production system as 'learning bureaucracy' and 'democratic Taylorism,' which on the surface is a contradiction of sorts. We, too, would like to refer to the contradictory nature of Hyundai Motor's production system as 'authoritative experimentalism.' This word is oxymoronic in expression, since experimentalism is mostly regarded as fundamentally democratic and non-authoritarian (Sabel, 1996, 2006).

'Democratic experimentalism' is chiefly focused on the problem of uncertainty (Herrigel, 2020; Sabel & Simon, 2011). This is an attempt to accommodate constant revision and change in context, and is in line with John Dewey's pragmatist philosophy, combining empowerment and central coordination of performance evaluation.[12] Sabel (1994, 2006) has applied the argument of democratic experimentalism to Toyota's lean production system. This highlights the positive properties of the lean production system, namely team-based integration between concept and execution, integration between development and mass production, and decentralization. The concern here is the matching and continuous change between the ideas and actions of the members involved at the collective level under a democratic order. In the lean production system, each activity is generally performed autonomously after agreement and consultation.

As a large hierarchical corporation, there has been a backlash against the harsh and intensive labor process inherent to Toyota's lean production system, earlier mentioned. Thus, it is debatable whether the company's system can be considered truly democratic. In Williamson's (1975)

[12] Simon (2015) points out that an experimental organization configuration entails principles of transparency and inclusion, modifiable planning, and improvement-oriented self-surveillance. Herrigel (2020) points to this as a form of dynamic capability. Accordingly, building up dynamic capabilities is possible even in large hierarchical firms (Zollo & Winter, 2002). On the other hand, the dynamic capability of democratic experimentalism incentivizes organizations to self-organize recursive learning processes.

sense, large corporations are inherently hierarchical compared to the market, so it would be absurd to call the resulting bureaucratic system of decision-making democratic. In this context, the meaning of 'democratic' may be understood as referring to 'inclusive' participation and collective decision-making among members within certain structural constraints. Experimentalism, on the other hand, implies continuous innovation and improvement (Herrigel, 2020; Sabel, 1994).[13]

Toyota's lean production system is synonymous with democratic experimentalism due to the properties of the system (Sabel, 1994; Simon, 2015). In a lean production system, the goal of each team is initially provisional, but forms a routine through improvement activities from the shop-floor. This in turn builds dynamic capabilities through incremental improvements (Zollo & Winter, 2002), and the process is experimental and democratic in that it involves improvement and synchronization of the just-in-time process flow and is carried out through consultation and agreement among stakeholders, including shop-floor workers and external partners. Sabel (1994) refers to this constant interaction between actual performance and ways to improve as 'learning by monitoring.' In Toyota's lean production system, the boundaries of task allocation between individuals and groups are flexible, so it is easy to train multi-functional and multi-skilled workers through team-oriented job rotation. In contrast, boundaries between tasks are clearly set around individual Meisters in Germany, and team-based job rotation among them does not work as well since workers have a unique skill identity.

The ongoing discussion of team-based work boundaries and mutual accountability lies at the heart of experimentalism in the lean production system. In this way, concept and execution are integrated. On the other hand, the integration of concept and execution is only partial at Hyundai Motor, as this improvement process is replaced by task-force team activities in the pilot center or assisted by integration between process technology engineers and skilled workers (see Chapters 3 and 4 in detail).

[13] Herrigel (2020) cites hierarchical insulation, exclusion of stakeholders, and inappropriate empowerment for participants as obstacles to democratic experimentalism in organizations such as large corporations. Democratic experimentalism, which involves iterative learning and restructuring of existing organizational boundaries, is a complex mixture of shared problem-solving, team-based goal setting, hierarchical protection strategies, and stakeholders' inclusion and exclusion in reality (Herrigel, 2020). Therefore, it can be seen that there are various forms in reality.

This means Hyundai Motor's situation cannot be viewed as fully democratic because it involves problem-solving through partial participation and deliberation rather than full participation and shared responsibility among members. The head of the *chaebol* also only grants autonomy to the engineer group and sets abstract goals somewhat unilaterally, which represents an authoritarian element. In spite of these structural limitations, Hyundai Motor has attempted to build dynamic capabilities by adopting experimentalist practices such as granting the aforementioned autonomy, problem-solving through improvisational learning at the pilot center, and inviting participation from various stakeholders.

In summary, experimentalism refers to one way of building up dynamic capabilities. This mainly occurs in a turbulent and uncertain environment, organizations with a strong learning culture, and when performing ambitious tasks (Zollo & Winter, 2002). While the first applies to both Toyota and Hyundai Motor, the second is more connected to Toyota, and the third to Hyundai Motor. As an evolutionary learning firm, Toyota's dynamic capabilities have been built up in a cumulative and sequential manner from the shop-floor, while Hyundai Motor has boldly accommodated post-Chandler practices by utilizing Chandlerian resources on top of a pre-Chandlerian ownership structure to develop in an agile and leapfrogging fashion. This is why if Toyota's building of dynamic capabilities can be dubbed 'democratic experimentalism,' we believe that Hyundai Motor deserves to be called 'authoritarian experimentalism.'

1.3.2.3 Hyundai Motor's Agile Production System as an Authoritarian Experimentalism

In this book, we introduce the concept of Hyundai Motor's production system as an agile production system in which the organizing principle of authoritarian experimentalism has been embedded. Hyundai Motor has successfully achieved agility in the midst of unpredictable and complex environmental changes in the twenty-first century, in a way that differs from Toyota's lean production system. The agile production system implemented by Hyundai Motor can be seen as the company's unique approach, influenced by the agile production system developed by American automakers.

Hyundai Motor's agility is attributed to the bold leadership of *chaebol* executives, the rapid and improvisational project execution by engineers, the preference for technocratic solutions, and the extended quasi-vertical integration of parts suppliers. Notably, Hyundai Motor's agile production system differs from Toyota's lean production system in several ways. The following chapters will specifically address the unique characteristics of this production system at Hyundai Motor.

References

Abernathy, W. J. (1978). *The Productivity Dilemma*. Johns Hopkins University Press.

Abo, T. (1994). *Hybrid Factory: The Japanese Production System in the United States*. Oxford University Press.

Abramovitz, M. (1986). Catching Up, Forging Ahead, and Falling Behind. *The Journal of Economic History, 46*(2), 385–406.

Acemoglu, D., Aghion, P., & Zilibotti, F. (2006). Distance to Frontier, Selection, and Economic Growth. *Journal of the European Economic Association, 4*(1), 37–74.

Adler, P. S., & Cole, R. E. (1993). Designed for Learning: A Tale of Two Auto Plants. *Sloan Management Review, 34*(3), 85–94.

Aghion, P., Guriev, S., & Jo, K. (2021). *Chaebols* and Firm Dynamics in Korea. *Economic Policy, 36*(108), 593–626.

Alchian, A. A., & Demsetz, H. (1972). Production, Information Costs, and Economic Organization. *The American Economic Review, 62*(5), 777–795.

Alinsky, S. D. (1969). *Rules for Radicals*. Random House.

Amsden, A. (1989). *Asia's Next Giant: South Korea and Late Industrialization*. Oxford University Press.

Amsden, A. (2001). *The Rise of the Rest: Challenges to the West from Late-Industrializing Economies*. Oxford University Press.

Amsden, A. H., & Hikino, T. (1994). Project Execution Capability, Organizational Know-How and Conglomerate Corporate Growth in Late Industrialization. *Industrial and Corporate Change, 3*(1), 111–147.

Aoki, M. (2001). *Toward a Comparative Institutional Analysis*. MIT Press.

Aoki, M. (2006). *Whither Japan's Corporate Governance?* (SIEPR Discussion Paper, 05-14). Stanford Institute for Economic Policy Research, Stanford University.

Aoki, M. (2010). *Corporations in Evolving Diversity: Cognition, Governance, and Institutions*. Oxford University Press.

Aoki, M., Kim, H.-K., & Okuno, M. (1997). *The Role of Government in East Asian Economic Development: Comparative Institutional Analysis*. Clarendon Press.

Babson, S. (1995). Lean Production and Labor: Empowerment and Exploitation. In S. Babson (Ed.), *Lean Work: Empowerment and Exploitation in the Global Auto Industry* (pp. 1–37). Wayne State University Press.

Becattini, G. (2002). From Marshall's to the Italian "Industrial Districts": A Brief Critical Reconstruction. In A. Q. Curzio & M. Fortis (Eds.), *Complexity and Industrial Clusters: Dynamics and Models in Theory and Practice* (pp. 83–106). Physica-Verlag.

Becker, G. S. (1993). *Human Capital: A Theoretical and Empirical Analysis with Special Reference to Education*. University of Chicago Press.

Berle, A., & Means, G. (1932). *The Modern Corporation and Private Property*. Commerce Clearing House.

Boyer, R., & Freyssenet, M. (1995). L'émergence de Nouveaux Modèles Industriels. *Actes du GERPISA, 15*, 75–142.

Boyer, R., & Freyssenet, M. (2002). *The Productive Models: The Conditions of Profitability*. Palgrave.

Braverman, H. (1974). *Labor and Monopoly Capital: The Degradation of Work in the Twentieth Century*. Monthly Review.

Brusco, S. (1992). Small Firms and the Provision of Real Services. In F. Pyke & W. Sengenberger (Eds.), *Industrial Districts and Local Economic Regeneration* (pp. 177–196). International Institute for Labour Studies, ILO.

Busemeyer, M. R. (2009). Asset Specificity, Institutional Complementarities and the Variety of Skill Regimes in Coordinated Market Economies. *Socio-Economic Review, 7*(3), 375–406.

Chandler, A. D. (1977). *The Visible Hand: The Managerial Revolution in American Business*. Harvard University Press.

Chandler, A. D. (1990). *Scale and Scope: The Dynamics of Industrial Capitalism*. Harvard University Press.

Chang, H.-J., & Andreoni, A. (2020). Industrial Policy in the 21st Century. *Development and Change, 51*(2), 324–351.

Chang, S.-J. (2003). *Financial Crisis and Transformation of Korean Business Groups*. Cambridge University Press.

Chesbrough, H. W. (2003). *Open Innovation: The New Imperative for Creating and Profiting from Technology*. Harvard Business School Press.

Cho, S.-J. (2014). Success in Catch-Up and Post Catch-Up Problem: Focusing on Hyundai Motor Group Case. *Journal of Korean Social Trend and Perspective, 91*, 136–168. (In Korean).

Cho, S.-J., Jeong, J. H., & Hwang, S. (2008). *The Korean Economy and Changes in Employment and Labor Regime*. Korea Labor Institute.

Ciborra, C. U. (1999). Notes on Improvisation and Time in Organizations. *Accounting, Management and Information Technology, 9*(1), 77–94.

Clark, K. B., & Fujimoto, T. (1991). *Product Development Performance: Strategy, Organization, and Management in the World Auto Industry*. Harvard Business School Press.

Clarke, C. (2005). *Automotive: Production Systems and Standardisation: From Ford to the Case of Mercedes-Benz*. Physica-Verlag.

Coase, R. (1937). The Nature of the Firm. *Economica, 4*(16), 386–405.

Coriat, B., & Dosi, G. (1998). Learning How to Govern and Learning How to Solve Problems: On the Co-Evolution of Competences, Conflicts and Organizational Routines. In A. Chandler, P. Hagström, & Ö. Sölvell (Eds.), *The Dynamic Firm: The Role of Technology, Strategy, Organizations and Regions* (pp. 103–133). Oxford University Press.

Cyert, R., & March, J. G. (1963). *A Behavioural Theory of the Firm*. Prentice-Hall.

Di Maio, M. (2008). Industrial Policies in Developing Countries: History and Perspectives. Quaderno di Dipartimento No. 48, Dipartimento di Istituzioni Economiche e Finanziarie, Universita degli Studi di Macerata.

DiMaggio, P., & Powell, W. (1991). The Iron Cage Revisited: Institutional Isomorphism and Collective Rationality in Organizational Fields. *American Sociological Review, 48*(2), 147–160.

Dohse, K., Jürgens, U., & Nialsch, T. (1985). From "Fordism" to "Toyotism"? The Social Organization of the Labor Process in the Japanese Automobile Industry. *Politics & Society, 14*(2), 115–146.

Dore, R. (1992). Goodwill and the Spirit of Market Capitalism. In M. Granovetter & R. Swedberg (Eds.), *The Sociology of Economic Life* (pp. 159–180). Westview Press.

Dosi, G. (1988). The Nature of the Innovation Process. In G. Dosi, C. Freeman, R. Nelson, G. Silverberg, & L. Soete (Eds.), *Technical Change and Economic Theory* (pp. 221–238). Pinter.

Dosi, G., Marsili, O., Orsenigo, L., & Salvatore, R. (1995). Learning, Market Selection and the Evolution of Industrial Structures. *Small Business Economics, 7*(6), 411–436.

Dosi, G., Nelson, R. R., & Winter, S. (2000). *The Nature and Dynamics of Organizational Capabilities*. Oxford University Press.

Dougherty, D. (1992). Interpretive Barriers to Successful Product Innovation in Large Firms. *Organization Science, 3*(2), 179–202.

Elbaum, B., & Lazonick, W. (Eds.). (1986). *The Decline of the British Economy*. Oxford University Press.

Estévez-Abe, M., Iversen, T., & Soskice, D. (2001). Social Protection and the Formation of Skills. In P. A. Hall & D. Soskice (Eds.), *Varieties of Capitalism: The Institutional Foundations of Comparative Advantage* (pp. 145–183). Oxford University Press.

Evans, P. (1995). *Embedded Autonomy: States and Industrial Transformation*. Princeton University Press.

Evans, P. (2021). Alice Amsden: A Reasoning Revolutionary in Development Economics. *Development and Change, 52*(4), 988–1008.

Fagerberg, J. (2005). Innovation: A Guide to the Literature. In J. Fagerberg, D. C. Mowery, & R. R. Nelson (Eds.), *Oxford Handbook of Innovation* (pp. 1–26). Oxford University Press.

Fagerberg, J., & Godinho, M. (2005). Innovation and Catch-Up. In J. Fagerberg, D. C. Mowery, & R. R. Nelson (Eds.), *The Oxford Handbook of Innovation* (pp. 514–542). Oxford University Press.

Freyssenet, M. (2005). Trajectoires Nationales et Trajectoires de Firmes: Esquisse d'un Schema d'Analyse. *Actes du GERPISA, 38*, 25–61.

Freyssenet, M., Mair, A., Shimizu, K., & Volpato, G. (Eds.). (1998). *One Best Way? Trajectories and Industrial Models of the World's Automobile Producers*. Oxford University Press.

Fujimoto, T. (1999). *The Evolution of a Manufacturing System at Toyota*. Oxford University Press.

Fujimoto, T. (2006). Architecture-Based Comparative Advantage in Japan and Asia. In K. Ohno & T. Fujimoto (Eds.), *Industrialization of Developing Countries: Analysis by Japanese Economics* (pp. 1–10). National Graduate Institute of Policy Studies.

Fujimoto, T. (2014). Auto Industry Life Cycle. *Journal of Product Innovation Management, 31*(1), 8–16.

Gerschenkron, A. (1962). *Economic Backwardness in Historical Perspective*. Harvard University Press.

Gill, I., & Kharas, H. (2007). *An East Asian Renaissance: Ideas for Economic Growth*. The World Bank.

Gong, Y. T., Baker, T., & Miner, A. S. (2006). *Capabilities and Routines in New Organizations: Evidence from the Field* (Working Paper). School of Business, University of Wisconsin—Madison.

Granovetter, M. S. (1973). The Strength of Weak Ties. *American Journal of Sociology, 78*(6), 1360–1380.

Guillen, M. (2001). *The Limits of Convergence: Globalization and Organizational Change in Argentina, South Korea, and Spain*. Princeton University Press.

Hall, P. A., & Soskice, D. (2001). *Varieties of Capitalism: The Institutional Foundations of Comparative Advantage*. Oxford University Press.

Hampson, I. (1999). Lean Production and the Toyota Production System Or, the Case of the Forgotten Production Concepts. *Economic and Industrial Democracy, 20*(3), 369–391.

Harvey, D. (2003). *The New Imperialism*. Oxford University Press.

Hattori, T. (1986). Technology Transfer and Management Systems. *The Developing Economies, 24*(4), 314–325.

Hattori, T. (2005). *Economic Sociology of Development: Korea's Economic Development and Social Changes*. Bunshindo Publishing. (In Japanese)

Helfat, C. E., & Peteraf, M. A. (2003). The Dynamic Resource-Based View: Capability Lifecycles. *Strategic Management Journal, 24*(10), 997–1010.

Helfat, C. E., & Winter, S. G. (2011). Untangling Dynamic and Operational Capabilities: Strategy for the (N)ever-Changing World. *Strategic Management Journal, 32*(11), 1243–1250.

Helper, S., MacDuffe, J. P., & Sabel, C. F. (2000). Pragmatic Collaborations: Advancing Knowledge While Controlling Opportunism. *Industrial and Corporate Change, 9*(3), 443–488.

Herrigel, G. (2000). *Industrial Constructions: The Sources of German Industrial Power*. Cambridge University Press.

Herrigel, G. (2004). Emerging Strategies and Forms of Governance in High-Wage Component Manufacturing Regions. *Industry and Innovation, 11*(1–2), 45–79.

Herrigel, G. (2010). *Manufacturing Possibilities: Creative Action and Industrial Recomposition in the US, Germany and Japan*. Oxford University Press.

Herrigel, G. (2020). Experimentalist Systems in Manufacturing Multinationals: Recursivity and Continuous Learning through Destabilization. In J. Glückler, G. Herrigel, & M. Handke (Eds.), *Knowledge for Governance: Knowledge and Space* (Vol. 15, pp. 415–439). Springer.

Herrigel, G., & Sabel, C. F. (1999). Craft Production in Crisis: Industrial Restructuring in Germany During the 1990s. In P. D. Culpepper & D. Finegold (Eds.), *The German Skills Machine: Sustaining Comparative Advantage in a Global Economy* (pp. 77–114). Berghahn Book.

Highsmith, J. (2004). *Agile Project Management: Creating Innovative Projects*. Addison Wesley.

Janoski, T., & Lepadatu, D. (2021). *The Cambridge International Handbook of Lean Production: Diverging Theories and New Industries Around the World*. Cambridge University Press.

Jensen, M. C., & Meckling, W. H. (1976). Theory of the Firm: Managerial Behaviour, Agency Costs and Ownership Structure. *Journal of Financial Economics, 3*(4), 305–360.

Jeong, J. H. (2021). The History of Engineer Training Policy as Part of Technical Manpower Training Policy. In S.-J. Cho (Ed.), *Labor Force Utilization Structure and Development Issues in the Korean Manufacturing Industry* (pp. 154–217). Korea Labor Institute. (In Korean)

Jeong, J. H. (2022). Evaluation and Challenges of Korea's 'Industrial System' for Co-prosperity and Solidarity. In K.-Y. Shin & H. Yoon (Eds.), *Korea Trapped in Success: The Political-Economic Trajectory of Regressive and Selective Welfare* (pp. 69–138). Seoul: Humanitas. (In Korean)

Jo, H. J. (1992). *A Study on the Production System of the Korean Automotive Industry: Focusing on Changes in the International Division of Labor in the 1980s*. Unpublished Doctoral Dissertation, Seoul National University. (In Korean)

Jo, H. J. (2016). *Hyundai Motor Company's Agile Production System: An Exploration of Korean Production System*. Hanul. (In Korean)

Johnson, C. (1982). *MITI and the Japanese Miracle: The Growth of Industrial Policy, 1925–1975*. Stanford University Press.

Johnson, C. (1999). The Developmental State: Odyssey of a Concept. In M. Woo-Cummings (Ed.), *The Developmental State* (pp. 32–60). Cornell University Press.

Joo, M. (2007). *A Study on the 'Japanization' of the Korean Automotive Industry Production System*. Unpublished Doctoral Dissertation, Kyungpook National University. (In Korean)

Jung, S. G. (1995). The Politics of Flexible Production and Work Reorganization: The Case Study of Hyundai Motor Company. *Economy and Society, 28*, 219–245. (In Korean)

Juravich, T. (1985). *Chaos on the Shop Floor: A Worker's View of Quality, Productivity, and Management*. Temple University Press.

Jürgens, U. (1991). Departures from Taylorism and Fordism: New Forms of Work in the Automobile Industry. In B. Jessop, H. Kastendiek, K. Nielsen, & O. Pedersen (Eds.), *The Politics of Flexibility: Restructuring State and Industry in Britain, Germany and Scandinavia* (pp. 233–247). Edward Elgar.

Jürgens, U. (1992). Internationalization Strategies of Japanese and German Automobile Companies. In S. Tokunaga, N. Altmann, & H. Demes (Eds.), *New Impacts on Industrial Relations: Internationalization and Changing Production Strategies* (pp. 63–96). Iudicium Verlag.

Jürgens, U. (1994). Lean Production. In H. Corsten (Ed.), *Handbuch Produktionsmanagement* (pp. 369–379). Gabler.

Jürgens, U. (1995). Lean Production and Co-Determination: The German Experience. In S. Babson (Ed.), *Lean Work: Empowerment and Exploitation in the Global Auto Industry* (pp. 292–308). Wayne State University Press.

Jürgens, U. (1998). Implanting Change: The Role of 'Indigenous Transplants' in Transforming the German Productive Model. In R. Boyer, E. Charron, U. Jürgens, & S. Tolliday (Eds.), *Between Imitation and Innovation, the Transfer and Hybridisation of Productive Models in the International Automobile Industry* (pp. 319–341). Oxford University Press.

Jürgens, U. (2002). Corporate governance, innovation, and economic performance: a case study on Volkswagen [Veröffentlichungsreihe/Wissenschaftszentrum Berlin für Sozialforschung, Forschungsschwerpunkt Technik-Arbeit-Umwelt, Abteilung Regulierung von Arbeit, 02-205]. Wissenschaftszentrum Berlin für Sozialforschung gGmbH.

Jürgens, U., & Krzywdzinski, M. (2016). *The New Worlds of Work: Varieties of Work in Car Factories in the BRIC Countries*. Oxford University Press.

Kenney, M., & Florida, R. (1988). Beyond Mass Production: Production and the Labor Process in Japan. *Politics & Society, 16*(1), 121–158.

Keynes, J. M. (1930). *A Treatise on Money, The Collected Writings of John Maynard Keynes* (Volume 5). Cambridge University Press.

Khan, M. (2000). Rent-Seeking as Process. In M. Khan & K. S. Jomo (Eds.), *Rents, Rent-Seeking and Economic Development: Theory and Evidence in Asia* (pp. 70 144). Cambridge University Press.

Khan, M., & Blankenburg, S. (2009). The Political Economy of Industrial Policy in Asia and Latin America. In G. Dosi, M. Cimoli, & J. E. Stiglitz (Eds.), *Industrial Policy and Development: The Political Economy of Capabilities Accumulation* (pp. 336–377). Oxford University Press.

Khanna, T., & Yafeh, Y. (2007). Business Groups in Emerging Markets: Paragons or Parasites? *Journal of Economic Literature, 45*(2), 331–372.

Kidd, P. T. (1994). *Agile Manufacturing: Forging New Frontiers*. Addison-Wesley Publishing.

Kim, C. (2010). Modularization and Value Chain Reorganization: Korean Automotive Industry. *Korean Journal of Labor Studies, 16*(1), 235–273.

Kim, C., Jo, H. J., & Ho, J. J. (2011). Modular Production and Hyundai Production System: The Case of Hyundai MOBIS. *Economy and Society, 92*, 351–385.

Kim, J. (2003). The Ownership and Governance of Korean *Cheabols*. *Journal of Korean Economic Development, 9*(2), 77–95. (In Korean)

Kim, L. (1980). Stages of Development of Industrial Technology in a Less Developed Country: A Model. *Research Policy, 9*(3), 254–277.

Kim, L. (1997). *Imitation to Innovation: The Dynamics of Korea's Technological Learning*. Harvard Business School Press.

Kim, L., & Nelson, R. (2000). *Technology, Learning, and Innovation: Experiences of Newly Industrializing Economies*. Cambridge University Press.

Knight, F. H. (1921). *Risk, Uncertainty, and Profit*. Houghton Mifflin.

Langlois, R. N. (2002). The Capabilities of Industrial Capitalism. *Critical Review, 5*(4), 513–530.

Langlois, R. N. (2003). The Vanishing Hand: The Changing Dynamics of Industrial Capitalism. *Industrial and Corporate Change, 12*(2), 351–385.

Lazonick, W. (1990). Organizational Capabilities in American Industry: The Rise and Decline of Managerial Capitalism. *Business and Economic History, 19*, 35–54.

Lazonick, W. (2003). The Theory of the Market Economy and the Social Foundations of Innovative Enterprise. *Economic and Industrial Democracy, 24*(1), 9–44.

Lazonick, W. (2005). The Innovative Firm. In J. Fagerberg, D. C. Mowery, & R. R. Nelson (Eds.), *Oxford Handbook of Innovation* (pp. 29–55). Oxford University Press.

Lazonick, W. (2010). The Chandlerian Corporation and the Theory of Innovative Enterprise. *Industrial and Corporate Change, 19*(2), 317–349.

Lazonick, W. (2015). When Managerial Capitalism Embraced Shareholder-Value Ideology: Comments on Duménil and Lévy. *International Journal of Political Economy, 44*(2), 90–99.

Lazonick, W., & O'Sullivan, M. (2000). Maximizing Shareholder Value: A New Ideology for Corporate Governance. *Economy and Society, 29*(1), 13–35.

Lee, B. C. (2013). Two-Pronged Structuration of Ownership, Control and Varieties of Capitalism: Based on the Critical Reformulation of the Theory of the Corporate Capitalism. *Journal of Korean Social Trend and Perspective, 89*, 222–258. (In Korean)

Lee, B. C. (2020). *The Making of Korean Capitalism: The Duet of Compression and Imbalance*. Haenam. (In Korean)

Lee, B.-H., & Jo, H.-J. (2007). The Mutation of the Toyota Production System: Adapting the TPS at Hyundai Motor Company. *International Journal of Production Research, 45*(16), 3665–3679.

Lee, K. (2013). *Schumpeterian Analysis of Economic Catch-up: Knowledge, Path-Creation, and the Middle-Income Trap*. Cambridge University Press.

Lee, K., & Lim, C. (2001). Technological Regimes, Catching-Up and Leapfrogging: The Findings from Korean Industries. *Research Policy, 30*(3), 459–483.

Lee, K., & He, X. (2009). The Capability of the Samsung Group in Project Execution and Vertical Integration: Created in Korea, Replicated in China. *Asian Business & Management, 8*(3), 277–299.

Lee, Y. H. (1994). *Fordism and Post-Fordism: A Comparison of Hyundai, Toyota and Volvo Car Plants*. Hanul. (In Korean)

Lepadatu, D., & Janoski, T. (2018). Just-In-Time Workforce? Temporary Workers as a Structural Aspect of Lean Production in the US Auto Industry. *International Journal of Automotive Technology and Management, 18*(2), 160–177.

Levy, B., & Kuo, W.-J. (1991). The Strategic Orientations of Firms and the Performance of Korea and Taiwan in Frontier Industries: Lessons from Comparative Case Studies of Keyboard and Personal Computer Assembly. *World Development, 19*(4), 363–374.

Lim, W. (2003). The Emergence of the *Chaebol* and the Origin of the *Chaebol* Problem. In S. Haggard, W. Lim, & E. Kim (Eds.), *Economic Crisis and Corporate Restructuring in Korea: Reforming the Chaebol* (pp. 35–52). Cambridge University Press.

Lim, W. (2012). *Chaebol* and Industrial Policy in Korea. *Asian Economic Policy Review, 7*(1), 69–86.

Lung, Y. (2005). The Link between the Diversity of Productive Models and the Variety of Capitalisms: A Review of the Literature and Contextualisation Using the Car Industry as a Case Study. *Actes du GERPISA, 38*, 109–140.

Marshall, A. (1890). *The Principles of Economics*. Macmillan.
Marshall, A. (1919). *Industry and Trade: A Study of Industrial Technique and Business Organization and of Their Influences on the Condition of Various Classes and Nations*. Macmillan.
Miner, A. S., Bassoff, P., & Moorman, C. (2001). Organizational Improvisation and Learning: A Field Study. *Administrative Science Quarterly, 46*(2), 304–337.
Myrdal, G. (1968). *An Asian Drama: An Inquiry into the Poverty of Nations*. Twentieth Century Fund and Pantheon Books.
Nelson, R., & Winter, S. (1982). *An Evolutionary Theory of Economic Change*. Harvard University Press.
Nonaka, I. (1990). Redundant, Overlapping Organization: A Japanese Approach to Managing the Innovation Process. *California Management Review, 32*(3), 27–38.
OECD. (2019). The Future of Doing Business. International Conference on Income-Led Growth, 3rd December, 2019, Seoul, Korea.
Ohno, T. (1988). *Toyota Production System: Beyond Large Scale Production*. Productivity Press.
Okazaki, T. (1999). Corporate Governance. In T. Okazaki & M. Okuno-Fujiwara (Eds.), *The Japanese Economic System and Its Historical Origins* (pp. 97–144). Oxford University Press.
Park, K., & Lee, K. (2006). Linking Technological Regimes and Technological Catch-Up: Analysis of Korea and Taiwan Using the US Patent Data. *Industrial and Corporate Change, 15*(4), 715–53.
Penrose, E. (1959). *The Theory of the Growth of the Firm*. Basil Blackwell.
Pérez, C., & Soete, L. (1988). Catching Up in Technology: Entry Barriers and Windows of Opportunity. In G. Dosi, C. Freeman, R. Nelson, G. Silverberg, & L. Soete (Eds.), *Technical Change and Economic Theory* (pp. 458–479). Pinter Publishers.
Piore, M. J., & Sabel, C. F. (1984). *The Second Industrial Divide: Possibilities for Prosperity*. Basic Books.
Prendergast, A. (1993). Marshallian External Economies. *The Economic Journal, 103*(417), 454–458.
Romme, G., Zollo, M., & Berends, P. (2010). Dynamic Capabilities, Deliberate Learning and Environmental Dynamism: A Simulation Model. *Industrial and Corporate Change, 19*(4), 1271–1299.
Sabel, C. F. (1994). Learning by Monitoring: The Institutions of Economic Development. In N. Smelser & R. Swedberg (Eds.), *Handbook of Economic Sociology* (pp. 137–165). Princeton University Press and Russell Sage Foundation.
Sabel, C. F. (1996). Learning-by-Monitoring: The Dilemmas of Regional Economic Policy in Europe. In OECD (Ed.), *Networks of Enterprises and*

Local Development (pp. 23–51). Organization for Economic Co-operation and Development.
Sabel, C. F. (2006). A Real-Time Revolution in Routines. In C. Heckscher & P. S. Adler (Eds.), *The Firm as a Collaborative Community* (pp. 106–156). Oxford University Press.
Sabel, C. F. (2007). Bootstrapping Development: Rethinking the Role of Public Intervention in Promoting Growth. In V. Nee & R. Swedberg (Eds.), *On Capitalism* (pp. 305–342). Stanford University Press.
Sabel, C. F., & Simon, W. H. (2011). Minimalism and Experimentalism in the Administrative State. *The Georgetown Law Journal, 100*(53), 53–93.
Sabel, C. F., & Zeitlin, J. (Eds.). (1997). *World of Possibilities: Flexibility and Mass Production in Western Industrialization.* Cambridge University Press.
Sabel, C. F., & Zeitlin, J. (2012). Experimentalist Governance. In D. Levi-Faur (Ed.), *The Oxford Handbook of Governance* (pp. 169–184). Oxford University Press.
Sako, M. (2003). Modularity and Outsourcing: The Nature of Co-evolution of Product Architecture and Organisation Architecture in the Global Automotive Industry. In A. Prencipe, A. Davies, & M. Hobday (Eds.), *The Business of Systems Integration* (pp. 229–253). Oxford University Press.
Sandberg, Å. (Ed.). (1994). *Enriching Production: Perspectives on Volvo's Uddevalla Plant as an Alternative to Lean Production.* Ashgate Publishing Limited.
Schmitz, H. (1999). Collective Efficiency and Increasing Returns. *Cambridge Journal of Economics, 23*(4), 465–483.
Schumpeter, J. A. (1909). On the Concept of Social Value. *The Quarterly Journal of Economics, 23*(2), 213–232.
Schumpeter, J. A. (1928). The Instability of Capitalism. *The Economic Journal, 38*(151), 361–386.
Schumpeter, J. A. (1934). *The Theory of Economic Development: An Inquiry into Profits, Capital, Credit, Interest, and the Business Cycle.* Harvard University Press.
Schumpeter, J. A. (1942). *Capitalism, Socialism and Democracy.* Harper and Row.
Schumpeter, J. A. (1954). *History of Economic Analysis.* Oxford University Pres.
Selwyn, B. (2011). Trotsky, Gerschenkron and the Political Economy of Late Capitalist Development. *Economy and Society, 40*(3), 421–450.
Sharp, J. M., Irani, Z., & Desai, S. (1999). Working Towards Agile Manufacturing in the UK Industry. *International Journal of Production Economics, 62*(1–2), 155–169.
Shimizu, K. (1995). Humanization of the Production System and Work at Toyota Motor Co. and Toyota Motor Kyushu. In Å. Sandberg (Ed.), *Enriching Production: Perspectives on Volvo's Uddevalla Plant as an Alternative to Lean Production* (pp. 383–404). Ashgate Publishing Limited.

Shiozawa, Y., Nishiyama, K., Yoshida K., Sato, O., & Tasaka, H. (1997). *Lectures in Complex Systems*. Diamond. (In Japanese)
Simon, W. H. (2015). The Organizational Premises of Administrative Law. *Law and Contemporary Problems, 78*(1–2), 61–100.
Skinner, W. (1985). *Manufacturing: The Formidable Competitive Weapon*. Wiley.
Song, H. K. (1999). *Labour Unions in the Republic of Korea: Challenge and Choice*. Discussion Papers (DP/107/1999), Labour and Society Programme, International Institute for Labour Studies, Geneva.
Song, W.-G. (2007). *Chaebol* System and the Corporate Social Responsibility. *Journal of Korean Social Trend and Perspective, 70*, 111–138. (In Korean)
Storper, M., & Salais, R. (1997). *Worlds of Production: The Action Frameworks of the Economy*. Harvard University Press.
Sturgeon, T. J. (2002). Modular Production Networks: A New American Model of Industrial Organization. *Industrial and Corporate Change, 11*(3), 451–496.
Teece, D. J. (1986). Profiting from Technological Innovation: Implications for Integration, Collaboration, Licensing, and Public Policy. *Research Policy, 15*(6), 295–305.
Teece, D. J. (2011). Human Capital, Capabilities and the Firm: Literati, Numerati, and Entrepreneurs in the 21st-century Enterprise. In A. Burton-Jones & J.-C. Spender (Eds.), *Oxford Handbook of Human Capital* (pp. 527–562). Oxford.
Teece, D. J. (2017). Towards a Capability Theory of (Innovating) Firms: Implications for Management and Policy. *Cambridge Journal of Economics, 41*(3), 693–720.
Teece, D. J., Pisano, G., & Shuen, A. (1997). Dynamic Capabilities and Strategic Management. *Strategic Management Journal, 18*(7), 509–533.
Thelen, K. (2004). *How Institutions Evolve: The Political Economy of Skills in Germany, Britain, the United States, and Japan*. Cambridge University Press.
Unger, R. (2009). *The Left Alternative*. Verso.
Unger, R. (2019). *The Knowledge Economy*. Verso.
Veblen, T. (1915a). *Imperial Germany and the Industrial Revolution*. Macmillan.
Veblen, T. (1915b). The Opportunity of Japan. *The Journal of Race and Development, 6*(1), 23–38.
Vera, D., & Crossan, M. (2005). Improvisation and Innovative Performance in Teams. *Organization Science, 16*(3), 203–224.
Verspagen, B. (2005). Innovation and Economic Growth. In J. Fagerberg, D. C. Mowery, & R. R. Nelson (Eds.), *The Oxford Handbook of Innovation* (pp. 487–513). Oxford University Press.
Wade, R. (1990). *Governing the Market: Economic Theory and the Role of Government in East Asian Industrialization*. Princeton University Press.
Wade, R. (2018). The Developmental State: Dead or Alive? *Development and Change, 49*(2), 518–546.

Weick, K. E. (1993). The Collapse of Sensemaking in Organizations: The Mann Gulch Disaster. *Administrative Science Quarterly, 38*(4), 628–652.

Whitford, J., & Zeitlin, J. (2004). Governing Decentralized Production: Institutions, Public Policy, and the Prospects for Inter-firm Collaboration in US Manufacturing. *Industry and Innovation, 11*(1/2), 11–44.

Whittaker, D. H., Sturgeon, T. J., Okita, T., & Zhu, T. (2020). *Compressed Development: Time and Timing in Economic and Social Development*. Oxford University Press.

Whittington, R., & Mayer, M. (2000). *The European Corporation*. Oxford University Press.

Williamson, O. E. (1975). *Markets and Hierarchies*. The Free Press.

Winter, S. G. (2003). Understanding Dynamic Capabilities. *Strategic Management Journal, 24*(10), 991–995.

Womack, J. P., Jones, D. T., & Roos, D. (1990). *The Machine That Changed the World*. HarperPerennial.

Wong, J. (2004). The Adaptive Developmental State in East Asia. *Journal of East Asian Studies, 4*(3), 345–362.

Woo-Cummings, M. (Ed.). (1999). *The Developmental State*. Cornell University Press.

World Bank. (1993). *The East Asian Miracle: Economic Growth and Public Policy*. Oxford University Press.

Young, A. A. (1928). Increasing Returns and Economic Progress. *The Economic Journal, 38*(152), 527–542.

Zeitlin, J. (2008). The Historical Alternatives Approach. In G. Jones & J. Zeitlin (Eds.), *The Oxford Handbook of Business History* (pp. pp.120–140). Oxford University Press.

Zollo, M., & Winter, S. G. (2002). Deliberate Learning and the Evolution of Dynamic Capabilities. *Organization Science, 13*(3), 339–351.

CHAPTER 2

An Overview of Hyundai Motor's Production System

The analysis in this book is mainly focused on the period after the 1997 Asian financial crisis, when Chairman Chung Mong-Koo took control of Hyundai Motor in the early 2000s. The period prior to this is described only briefly for the sake of providing background information to help readers understand the period after the 2000s. This period is when Hyundai Motor took off as a middle-ranked carmaker by releasing its own model and upgrading the company's engine development capabilities as part of 'catch-up 1.0.' On the other hand, the period after the 2000s is when the company grew into one of the top five global players based on modularization and the concentrated abilities of engineers as part of 'catch-up 2.0.'

With this in mind, this chapter analyzes the key characteristics of Hyundai Motor's production system by comparing it with that of Toyota through financial statements and other data. Since the details of Hyundai Motor's production system will be dealt with in what follows, this chapter provides more of a broad outline.

H. J. Jo et al., *Agile Against Lean*,
https://doi.org/10.1007/978-981-99-2042-6_2

2.1 Hyundai Motor's Production System Before the 1997 Asian Financial Crisis

Founded in December 1967, Hyundai Motor began as a company that imported and assembled knockdown (KD) parts in a joint venture with Ford from November 1968. An attempt was made to modify Ford's Cortina model for sale in Korea, but the joint venture collapsed in 1973 when Hyundai Motor requested support from Ford to develop independent technological capabilities, but Ford demanded that Hyundai Motor remain a subordinate part of the multinational corporation. Hyundai Motor subsequently pursued its own model development strategy and succeeded in releasing the Pony in 1974, the company's first step down the path of in-house development. The Long-Term Automobile Industry Promotion Plan released by the Korean government in January 1974 was aimed at localizing parts and served to facilitate Hyundai Motor's transition from simple assembly to manufacturing. Diesel engine manufacturing by the UK firm Perkins, body design by Italdesign, and a technical partnership with Mitsubishi in the engine and transmission sector all played an important role in Hyundai Motor's development of the Pony. The strong determination of the *chaebol* head also played a central role in the success of this strategy as management provided continued support and granted engineers a relatively high level of autonomy (Hyundai Motor Company, 1987).

In terms of manufacturing design technology, Hyundai Motor did not imitate the technology of any specific carmaker. Instead, the company independently established a technology acquisition plan and explored various learning paths. For example, the facilities used in automobile plants were mainly imported from Europe, but the expertise involved in making actual products came from Japan, as did equipment such as molds, welding machines for assembly, and jigs. Accordingly, Hyundai Motor's production system was a hybrid of American, European, and Japanese technology from the beginning, which was absorbed and reinterpreted in a unique way (Cho, 2014). In this way, Hyundai Motor was able to follow the path of developing its own model through a tailored combination of technologies by importing technical prowess from several countries.

Hyundai Motor went through countless trials and errors in the process of piecing together this jigsaw puzzle of foreign technologies, but ultimately owes a great debt to Mitsubishi. Through close cooperation with

Mitsubishi, Hyundai learned the basic technologies necessary for automobile production such as tools, materials, and facilities. In the process of developing the Excel, which was planned as an export model in the mid-1980s, Hyundai Motor acquired technology related to mass production from Mitsubishi. This involved a production management system for improving the company's capacity to use the large-scale facilities necessary for mass production, systematic quality control, and cost reductions (Hyundai Motor Company, 1992). However, the nature of this cooperative relationship with Mitsubishi has changed over time. In the 1970s and 1980s, the Mitsubishi model platform was adopted in the process of developing a series of products such as small to mid-large vans and Sport Utility Vehicles (SUVs), but joint development had taken over by the 1990s. This signifies that Hyundai Motor's technological capabilities had gradually advanced over time (Yeo, 2021). In the early days, expertise in plant operations under the Japanese production system depended heavily on the guidance of engineers who formerly worked for Japanese automakers (Kang, 1986).

In the 1980s, Hyundai Motor continued its path of self-reliant development by persuading policy authorities who had demanded a joint venture with GM as part of the restructuring of the Korean auto industry at that time. By the mid-1980s, the company had established a mass production system by utilizing the duopoly structure of the automobile market. By linking profits from this monopolistic market position with investment, hundreds of thousands of units were exported to the United States each year, creating an incredible success story for Excel[1] in the North American market. In November 1984, Hyundai Motor founded the Mabuk-ri Research Center in Yongin, Gyeonggi-do, and expanded its research and development efforts in an attempt to localize the technology that went into core parts. As a result, Hyundai Motor succeeded in developing an electronically controlled Alpha engine, instead of a carburetor one, and transmission on its own in 1991 as a typical example of technological leapfrogging by skipping successive steps in the development of engines, making Hyundai Motor the only company in a developing country with the ability to design and manufacture its own

[1] In 1986, Fortune magazine selected the Excel as one of the ten hit products of the year.

engines (Hyundai Motor Company, 1992; Kim, 1994).[2] Around this time, Hyundai Motor virtually caught up with advanced carmakers in almost all facets of automotive technology, having built technological prowess in powertrains, including engines and transmissions, as well as in production technology, body design, and styling (Kim, 1997; Lee & Lim, 2001).

In the late 1980s, Hyundai Motor attempted to benchmark Toyota to introduce a lean production system, the standard for automobile production systems at that time. Hyundai Motor's exports to North America plummeted as its quality failed to support the hype surrounding the Excel in the late 1980s, but the company was saved by 'relative stagnation' which refers to the fact that the company's growth was not as bad as expected compared to the past due to continued growth in the Korean market at the time, despite of its severe difficulty with a dramatic drop in car exports to the USA resulting from poor quality (Jo, 1992). As such, this suggests that it would be difficult for Hyundai Motor to grow into a top global player in the era of diversified consumption only with the growth strategy of simply increasing volumes through exploiting existing Taylor-Fordist stocks without substantial vehicle quality improvement.

In the 1990s, Hyundai Motor built a new plant in Asan to benchmark elements of the lean production system but ran into some problems. The lean production system requires voluntary participation and dedication from shop-floor workers, but this was not the case for Hyundai Motor. In contrast to the accumulation of independent technical capabilities in overall automotive technology through Hyundai Motor's cooperative relationship with Mitsubishi, this difference in institutional conditions acted as a significant barrier to benchmarking Toyota's lean production system.

The characteristics of the Korean labor market are the major factors in understanding Hyundai Motor's production system. In the 1970s and 1980s, Korea had a seniority pay system and lifetime employment akin to that of Japan, but Hyundai Motor did not show much interest in the formation of firm-specific skills for production workers. This is because the competitive edge of the company's main products at this time came from low-priced, small and medium-sized cars, so Hyundai Motor was more interested in establishing a mass production system to

[2] Kim (1998) interpreted Hyundai Motor's development of the Alpha engine as a shift from 'learning by doing' to 'learning by research.'

achieve economies of scale for standardized products, and these products did not demand a high level of quality. Prior to Korea's democratization in 1987, Hyundai Motor was able to maintain low wages and harsh working conditions with the support of the authoritarian government in enforcing oppressive labor controls (Song, 1999).

Since the democratization of Korea and technological advancement of the late 1980s, Hyundai Motor has been called on to produce higher quality cars and respond flexibly to changes in market demand. Although the company attempted to introduce a 'vocational qualification system' to help production workers develop firm-specific skills, this ultimately failed (Jo, 2005). Hyundai Motor has also suffered from strikes almost every year since the establishment of the union in 1987. When confrontational labor–management relations made it difficult to expect positive effects from vocational training, Hyundai Motor gave up on the vocational qualification system. However, an internal labor market formed at the same time due to the increased bargaining power of the union and a shortage of labor. The union's deep mistrust of the company thwarted the flexible use of the workforce and performance-based HR management strategies.

Around 1990, Hyundai Motor began pursuing a strategy of flexible automation that largely excluded labor. This involved adopting the latest technology instead of actively utilizing workers due to confrontational labor relations (Lee & Jo, 2007).[3] Hyundai Motor sought to gain a competitive advantage in the potential of information and communications technology rather than shop-floor labor (Jung, 1995). To address the problems of just-in-time supply of parts and synchronizing production processes, these strategies extended beyond the boundaries of the firm. Investment in flexible automation continued throughout the 1990s, including the establishment of integrated production management (CIM) following the launch of the Local Area Network (LAN) and Wide Area Network (WAN) system in December 1988, improvement to the

[3] The Asan Plant, which was built to escape the confrontational labor–management relations of the Ulsan Plant, was completed in November 1996. The Asan Plant can be called a green-field plant in that it hired a large number of production workers who had no experience in union activities (Chung, 1997). However, the Asan plant's labor relations took on a nature similar to those of the Ulsan plant within a few years. Since workers belonging to the same company could easily communicate among themselves, confrontational labor–management relations spread to the Asan Plant.

Assembly Line Control (ALC) system based on the Distributed Automation Edition (DAE) system in September 1990, and the application of the Flexible Body Line (FBL) method (Jo, 2005).

Unlike other automakers in developing countries, Hyundai Motor has continued to develop strategies for developing its own models through a process of learning and absorbing different technologies by weaving them together. From the beginning, management and engineers were accustomed to entrepreneurial flexibility and pragmatic thinking. However, this was not the case for shop-floor labor. Hyundai Motor was unable to fully benchmark lean production systems that make it difficult for executives or engineers to exert discretion unlike large Chandlerian firms, due to the differences in social and institutional conditions that made it impossible to unleash the full potential of workers. Nor was the company able to implement the accumulation of manufacturing capabilities through routines as with Neo-Schumpeterian evolutionary learning firms.

2.2 A Macro-institutional Overview of Hyundai Motor's Production System After the 1997 Asian Financial Crisis

2.2.1 *The 'Middle-Ranked Carmaker's Trap'*

As described above, Hyundai Motor developed its own engines in the early 1990s, catching up with leading players in almost all areas of automotive technology. However, the success of the Excel in the late 1980s collapsed due to quality issues and problems with the after-sales maintenance network. From that point on, Hyundai Motor appeared as a regular butt of jokes related to 'low-quality and low-priced cars' on American TV shows until the end of the 1990s. In addition, the attempt to benchmark Toyota's lean production system encountered resistance due to hostile labor–management relations.

As shown in Fig. 2.1, except for a temporary rise in return on equity (ROE) from the late 1980s to the mid-1990s, profitability continued to deteriorate until the end of the 1990s due to intensified domestic competition and quality problems. Fortunately, this decline was offset due to the remarkable growth of the Korean domestic market (Jo, 1992). J.D. Power's Initial Quality Study (IQS) investigates quality by studying the frequency of problems per 100 vehicles over the first 90 days from purchase for vehicles less than one-year-old. Although Hyundai Motor's relative IQS index has been on a downward trend since the mid-1990s, it is difficult to view the company as an advanced carmaker as it remained above the industry average throughout the 1990s. In contrast, Toyota's relative IQS index remained below the industry average over the same period (see Fig. 2.1).

In this sense, Hyundai Motor succeeded in its initial attempt to catch up and be able to compete with advanced carmakers in almost all areas of automotive technology. However, since the company has failed to achieve sufficient efficiency in its own production system to steadily increase quality and profitability. As the recent discussion of the middle-income trap in relation to the Korean economy (e.g. Aghion et al., 2021) implies, Hyundai Motor succeeded in catching up in the early 1990s through an investment-led growth strategy based on a mass production system of importing and absorbing existing technologies. However, the second challenge of growing into a leading carmaker through the production of high-quality automobiles that can respond to different consumer tastes has proven more difficult.

Hyundai Motor's profitability also deteriorated due to the sharp decline in domestic sales following the 1997 Asian financial crisis, and the company posted a loss for the first time since the early 1980s. Hyundai Motor was heavily dependent on the domestic market at the time and was not recognized as a top automaker due to declining exports as a result of quality problems. The company sought to enter the North American market through the construction of a plant with an annual capacity of 100,000 units in Bromont, Canada in 1989, but subsequently pulled out in 1993 due to quality issues and the lack of a sufficient sales network.

The unsatisfactory state of labor–management relations after the establishment of the Hyundai Motor union in 1987 also discouraged workers from voluntarily participating in skill development, which served as an obstacle to benchmarking Toyota's lean production system. Mass layoffs in the late 1990s further strengthened this distrust between labor and management. When the capacity utilization rate dropped to 40% and the company faced a management crisis in 1997, 18,730 workers were made redundant and another 10,000 had their employment status 'adjusted' through voluntary retirement and layoffs. This led to the union making employment protection a top priority. Job insecurity created resistance to reorganizing the workforce within the company, making it virtually impossible to relocate employees even when transfers were required between plants due to changes in demand for a specific model.

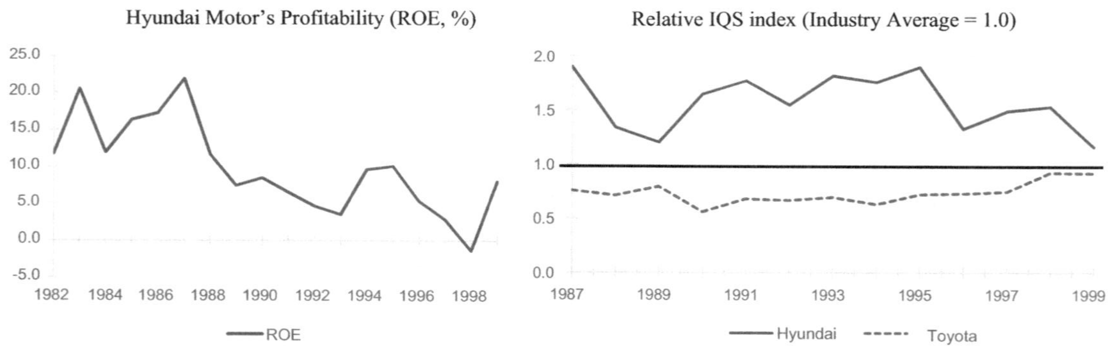

Fig. 2.1 Hyundai Motor's profitability and relative quality level in the 1980s and 1990s (*Note* Based on unconsolidated financial statements. *Source* KIS-VALUE [http://www.kisvalue.com], J.D. Power [https://www.jdpower.com], and author's own creation)

In summary, Hyundai Motor had fallen into the 'middle-ranked automaker's trap' by the 1990s due to quality problems, falling profitability, failed direct investment in Canada, and deepening distrust between management and the unions. At that time, the prevailing foreign view of the Korean automobile industry was gloomy. For example, Ravenhill (2003) predicted it would be difficult for Hyundai Motor to escape the image of low-priced, low-quality vehicles due to a lack of technological independence. After seeing Hyundai Motor flounder in the US market since the 1980s, Womack et al. (1990: 261–263) believed it would be impossible for Korea to successfully introduce a Japanese production system to gain a competitive advantage.

2.2.2 *Entrepreneurial Management Exploits 'Windows of Opportunity'*

Times of restructuring, fluctuations in demand, and organizational and technological changes are bound to open windows of opportunity. As discussed by Gerschenkron (1962), it is impossible to catch up to other companies through a sequential accumulation of capabilities alone. The only way to gain ground is to explore windows of opportunity that arise from industrial transformation, demand fluctuations, shifts in government policy, and organizational and technological changes.

In 1999, Hyundai Motor acquired Kia Motors, which had become insolvent during the 1997 Asian Financial Crisis, through debt relief and an equity swap from creditors. This made Hyundai Motor the world's tenth-largest automobile company. In September 2000, new chairman Chung Mong-Koo formed the Hyundai Motor Group as an amalgamation of Hyundai Motor and Kia Motors. This entity was separate and independent from the Hyundai Group. Chung's new 'GT-5 (Global Top-Five)' vision was to develop Hyundai Motor and Kia into the world's fifth-largest automobile company by 2010. Hyundai Motor was able to solidify its monopoly position in the domestic market and enjoy economies of scale through the acquisition of Kia Motors. The new goal appeared to follow from the global oligopoly theory that only five major carmakers could survive in the global automobile market of the future.

Following Hyundai Motor's acquisition of Kia, which was considered an unreasonable M&A at the time, there was a flood of large-scale mergers and acquisitions in the global automobile industry after the mid-1990s. For example, Renault took over the management rights of Nissan in 1998, while Daimler and Chrysler merged (in an alliance with Mitsubishi) in 1999, Ford also acquired Volvo in 1999, and GM and Fiat exchanged shares in 2000. These moves were aimed at reducing costs and expanding sales through the economies of scale that resulted from platform sharing and improvements to part sourcing systems. Unlike previous M&As, this round had trans-regional characteristics (Lung, 2001). At the time, the prevalence of the global oligopoly theory led automakers to believe that only the top five giants could survive through super-large M&As.

New chairman Chung Mong-Koo announced a new policy of 'quality management' to radically improve quality in order to increase Hyundai Motor's brand image beyond the mass export of low-cost compact cars. This led to the delegation of authority to middle managers and the R&D department, with an emphasis on growth and performance. In March 1999, Hyundai Motor and Kia established the Quality Headquarters, which integrated quality work in each sector. The Quality Control Room was established in December later that year to monitor on-site quality issues 24 hours a day and propose solutions when problems arise. In 2002, this was upgraded to the General Quality Headquarters, and the organization in charge of overseas quality was incorporated into this body in 2003. Hyundai Motor and Kia's quality innovation was then standardized into four steps—registration, classification, improvement, and verification—through the establishment of the Global Quality Management System in 2005.

Managers and engineers were given authority to improve quality and had the autonomy to set more specific goals, go through trial and error, and conduct experiments to achieve their goals. This was rooted in the CEO's abstract goal of quality management to make 'cheap and high-quality' cars.

As Herrigel (2020) notes that problem-solving and dynamic capability building through participation and deliberation among various stakeholders emerged as a pattern in German automakers, the delegation of authority in Hyundai Motor has served as the basis for the pragmatic experimentation of post-Chandlerian practices discussed in Chapter 1, despite *chaebols* being recognized as Chandlerian big firms based on the pre-Chandlerian ownership base with no separation of ownership and management.

The 1990s saw the incorporation of Russia and Eastern Europe into the capitalist world, the emergence of huge new markets such as China and India in Asia, and South American nations such as Mexico and Brazil drawing attention as emerging markets. In these countries, the term 'low level of motorization' describes a situation in which automobiles have not yet become widely accessible to the public and are not considered a necessity of life. However, as the population grows, this can lead to increased local investment and subsequently, expansion of the market. In other words, the future growth potential of these emerging markets and the existence of a cheap but skilled surplus labor force have led to increased local investment in order not only to supply products to the local market but also to re-export them to developed countries. As a result, foreign direct investment, global production, and cross-border trade have all increased in the global automotive industry, facilitated by trade and investment liberalization through World Trade Organization (WTO) agreements. Despite the increase in production in emerging markets, North America, Japan, and the EU still account for around half of the global automobile production, which is a substantial figure (Sturgeon et al., 2009).

At that time, Hyundai Motor had a very low level of globalization in production compared to sales. This is partly due to adopting a prudent approach to foreign local investment after the failure in Bromont in 1993. However, in order to cope with exchange rate fluctuations, trade frictions, and changes in local consumer tastes and overcome the limited Korean market to become a top-five global company, Hyundai Motor had no choice but to expand overseas production capacity as it sought to make inroads into overseas markets. Accordingly, since the mid- to late-1990s, Hyundai Motor and Kia have entered Brazil (1995), Turkey (1995), India (1996), China (2002), Slovakia (2007), Czech Republic (2009), Alabama, USA (2005), Georgia, USA (2006), and Russia (2010).

The past 50 years in the global automobile industry can be summarized as an era of 'rapid incremental innovation' in product development (Clark & Fujimoto, 1991). Product and process innovation did not significantly change the integrated architecture, technological changes were evolutionary, and functional improvements through an electronic control system (ECU) contributed to greater competitiveness. In this respect, innovation has not been disruptive (Christensen, 1997), so major players in developed countries have survived and there has not been a sharp decline in the number of components and parts. From a technical standpoint, vehicle performance characteristics such as noise, vibration, and handling are closely related to each other in the automotive industry, making it difficult to quantify the interactions between different elements before you try them. It is not easy to increase modularity across the value chain through standard setting and codification due to the structural characteristics of the integrated architecture in which minor modifications have an immediate impact on the overall quality of a product (Sturgeon et al., 2008).

In contrast, switching the supply of individual parts to modular sourcing is not very difficult. This refers to outsourcing the management of modular parts supply to a 0.5-tier large supplier (Sako & Murray, 1999; Sturgeon & Lester, 2004), where the ability to design complex parts and subsystems shifts from final automakers to parts suppliers. However, since this can result in a loss of control over the innovation process (Cohen & Levin, 1989), there is a greater need for joint design between both parties (Sturgeon et al., 2008) and close coordination between production and logistics is required (Humphrey, 2003).

In the 1990s, the German automobile industry was facing restructuring and decided to introduce some elements of Japan's lean production system, including the reorganization of supplier relations (Jürgens, 2003). Since the German economic recession in 1993, first-tier suppliers came to play a more important role. The modularization of parts supply was first achieved at Daimler Chrysler's Micro Compact Car (MCC)-Hambach plant in 1996 and Volkswagen's Mosel plant in 1997. Reducing internal reliance on parts saved work time and cut down the labor required for investment and assembly processes, thus reducing labor costs and improving the JIT system. As a result, overall costs were reduced, parts procurement became more centralized, and investment in new markets fell.

In the 1990s, Hyundai Motor fully benchmarked Volkswagen's parts supply modularization and Mercedes-Benz's pilot production line operation. While the attempt to benchmark the lean production system before the 1997 Asian Financial Crisis ended in failure, the German practice of preferring technology-oriented solutions from engineers helped Hyundai Motor escape from its productivity dilemma and begin the second stage of catching up to advanced automakers. Benchmarking these practices was relatively easy, as they did not require extensive technological learning or skill formation. It was simply a matter of corporate organization governance of reorganizing first-tier subcontractors and production capacity investment of constructing pilot production lines, and these could be attained with the discretionary power and market influence of *chaebol* organizations. In this context, Chung made the bold decision to establish a pilot center within the site of Namyang Research and Development Center in 2003. This facility functioned as a hub for improving quality issues through participation and deliberation among various stakeholders.

In this sense, Hyundai Motor's escape from the middle-ranked carmaker trap in the late 1990s was spurred by actively leveraging a 'window of opportunity' such as technological changes (modularization and the global value chain) and changes in market demand (globalization), as well as government intervention and regulatory changes (restructuring and *chaebol* reforms). Veblen (1915) argued that catching up in technological terms is governed by pecuniary factors and opportunities in new industries. From this, we can see that the capacity to mobilize financial resources to take advantage of new opportunities and the entrepreneurial capacity of Hyundai Motor's executives played an important role in the 'catch-up 2.0 period' when the company had already accumulated internal capabilities to a certain extent.

2.2.3 *Corporate Governance Structure and the* Chaebol *System*

One of the major controversies over corporate governance concerns the scope of stakeholders and the institutional configuration of checks and balances among them. Corporate governance can be viewed as the process of allocating resources and profits by assuming a multiple-agency relationship among various stakeholders (Aoki, 2001; O'Sullivan, 2002).[4] Previous discussions on corporate governance in the West have mainly focused on the relationship between shareholders and managers, whereas in Korea, the controlling majority shareholders and managers are virtually identical due to the *chaebol* structure of conglomerates, so the literature has focused on the relationships between different types of shareholders. For example, Chang (2004) and Kim (2006) respectively focus on checks

[4] Aoki (2010) focuses on institutional complementarity between business organizations and governance, and emphasizes 'residual rights of use' that focus on the utilization of corporate human assets rather than 'residual rights of control' over the ownership and control of physical corporate assets. Emphasis on the former is related to the post-World War II ownership structure reform of *zaibatsu* in the Japanese context, whereas emphasis on the latter reflects the emergence of minority shareholders and diversification of ownership structures in a Western context. Aoki (2010) emphasizes the control and use of managers and workers' intangible assets over shareholder ownership of physical assets.

and balances between domestic capital and foreign capital and major shareholders and minority shareholders. This has the effect of excluding other important stakeholders from the discussion, especially labor.

If we view corporate governance as a multiple-agency problem, there can be greatly varying interests among key members of the governance structure such as shareholders, executives, and workers. Shareholders seek short-term profits, while executives are concerned with long-term growth and control, and workers care about job security and skill-building (Jürgens, 2002). Because of this divergence, there is no single best way to balance competing interests, and cooperation can be obtained in a number of ways. For example, there may appear in reality some combinations between key different corporate interests such as manager–worker (e.g.: 'productivist democratic capitalism'), shareholder–manager (e.g.: 'finance-led neoliberalism'), and manager–shareholder (e.g.: 'productivist neoliberalism') (Lee, 2013).

From the viewpoint of the multiple agency problem, it is reasonable to assume that corporate governance is composed of three key stakeholders: workers, shareholders, and executives or managers and that corporations operate through conflict and coordination among different interests. In order for a company to maximize profits in the short term, accumulate human assets in the mid- to long-term, and achieve growth through the exercise of leadership in the long term, the interests of these three parties must be balanced. In the Korean context, since the agency problem between internally controlling key shareholders including *chongsu* and executives or managers is less serious due to the nature of the *chaebol* structure, we think the governance structure can be described as shown in Fig. 2.2, which assumes the interests of the *chongsu* and executives or managers can be expressed as one and the same. As mentioned above, the *chongsu* and executives or managers mainly focused on control and long-term growth called the *chongsu-holder value or control-holder value*, while external minority shareholders pursue short-term profits such as dividends and rising stock prices called *shareholder value*, and labor seeks long-term employment stability and skill formation called *work-holder value*.

For example, Jürgens (2002) sees the change in the governance structure of the Volkswagen Group in the 1990s as a combination of shareholder value and work-holder value. Jürgens argues that work-holder value played a role in keeping shareholder values in check based on Germany's own 'diversified quality production' system, thereby laying the foundation for profitability and growth. Although Volkswagen

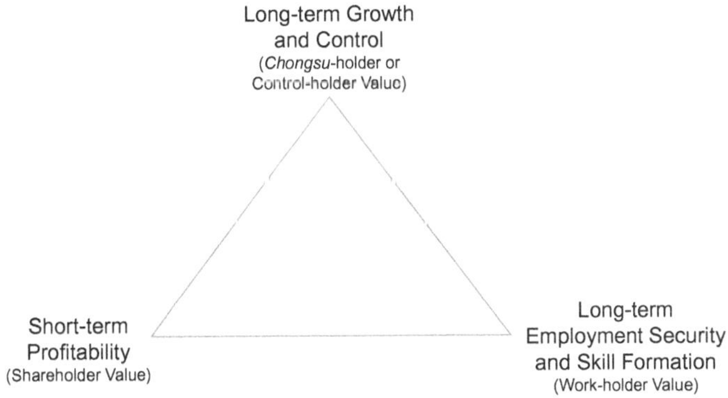

Fig. 2.2 Governance structure of *Chaebols* (*Source* Author's own adaptation from Jürgens, 2002)

followed the traditional German path of 'productivist democratic capitalism' through cooperation between management and labor without denying shareholder values, the company selected Ferdinand Piëch as CEO and accepted a certain degree of 'productivist neo-liberalism' in which management and shareholders cooperate, which can be seen as a hybrid of these forms of cooperation.

As of the end of 2021, the Hyundai Motor Group is a diversified business group of 57 affiliates including financial and non-financial subsidiaries and employs a workforce of 174,952 (see Fig. 2.3). Although the Hyundai Motor Group operates in numerous industrial sectors, it can be seen as a diversified business group based on the manufacturing industry, where it has 24 affiliates and employs 148,753 people. As shown in Fig. 2.4, complex equity relationships are formed with Hyundai Motor, Kia, Hyundai Mobis, Hyundai Glovis, and Hyundai Engineering and Construction at the center. This shows that related and unrelated diversifications based on quasi-vertical integration are mixed. In other words, *chaebols* are diversified business groups.

One notable feature of the ownership structure of *chaebols* is their identity as family businesses based on blood ties. Although the ownership of the *chongsu* family is small in the *chaebol* governance structure, there is a wide gap between ownership and control, since the *chongsu* family exercises control over the business group as a whole through shares in affiliates (Kim, 2003). As seen in Table 2.1, this is certainly true for Hyundai

Fig. 2.3 Hyundai Motor Group's subsidiaries and employment by industry (as of the end of 2021) (*Note* A [Agriculture, forestry, and fishing], C[Manufacturing]; D [Electricity, gas, steam, and air conditioning supply]; F [Construction]; G [Wholesale and retail trade]; H [Transportation and storage]; I [Accommodation and food services]; J [Information and communication]; K [Financial and insurance]; L [Real estate]; M [Professional, scientific, and technical activities]; N [Business facilities management and business support services; rental and leasing]; R [Arts, sports, and recreation-related services]. *Source* Korea Fair Trade Commission [2022] and author's own creation)

Motor Group. However, compared to the five largest *chaebol* conglomerates in Korea, Hyundai Motor's gap between ownership and control is smaller than that of Samsung Group.

It is possible for the *chongsu* to control affiliates with a minority stake through roundabout shareholding or cross-shareholding. Hyundai Motor Group has a roundabout shareholding structure that leads to 'Hyundai Mobis → Hyundai Motor → Kia Motors → Hyundai Mobis,' where the *chongsu* family controls the entire group with a minority stake (see Fig. 2.5). Core companies that play a key role in this structure include Hyundai Motor, Kia Motors, Hyundai Mobis, Hyundai Steel, and Hyundai Glovis.

A *chaebol* is an organizational structure that seeks collective efficiency. A closed production and innovation system operates based on quasi-vertical integration and strict ownership relationships. Economies of scale, specialization, knowledge embedded within the organization,

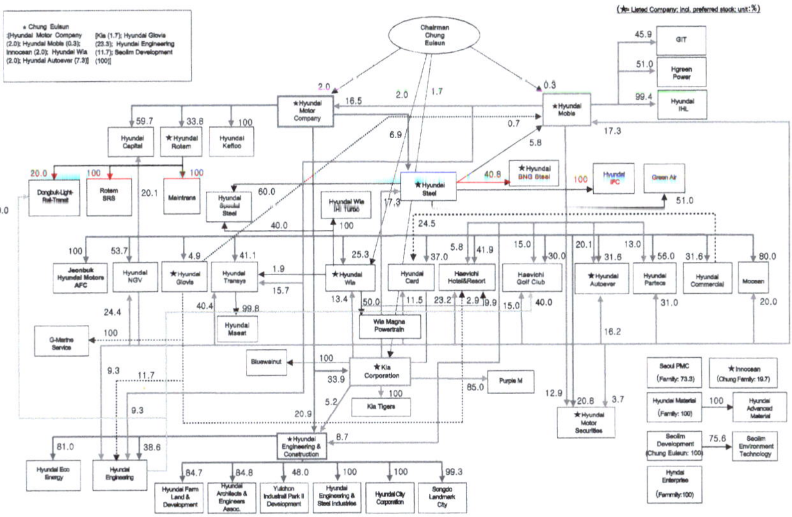

Fig. 2.4 Hyundai Motor Group's equity structure (as of the end of 2021) (*Source* Author's own adaptation from Korea Fair Trade Commission, 2022)

and business group-wide network externalities through related and unrelated diversification-based quasi-vertical integration serve to amplify relationship-specific synergies and enhance the competitiveness of the business group. However, this closed nature makes it difficult to adapt to changes in the external environment, which can cause tunnel vision and reduce competitiveness.

As business groups, *chaebols* have a governance structure that allows them to mobilize and allocate resources in the same way as individual enterprises. This can serve as the basis for organizational arrangements that create collective efficiency, but can also reveal the chaos of the governance structure caused by the tyranny of the head called *chongsu*. The key issue here is the problem of coordination, which has two dimensions. The first is coordination in terms of mobilizing and allocating resources to the right place in a timely way, and the second is mediation in terms of resolving the agency problem among various stakeholders such as the family of the head, executives or managers, and labor. The former is related to the mechanism of collective efficiency, while the latter is related

Table 2.1 Ownership structure of the five largest *Chaebols* (as of the end of 2021)

	Internal controlling shareholder's stake ratio (%)				*Sub-total* (%)	Chongsu's *real share ratio* (%)	*Control/Ownership*
	Chongsu *family* (*a*)		*Affiliates*	*Others*	*A*	$B = a/A$	$C = A/B$
Samsung	1.5	(0.5)	41.6	2.6	45.6	3.3	13.8
Hyundai Motor	6.3	(2.6)	47.5	1.5	55.2	11.4	4.8
SK	1.0	(0.5)	56.7	3.2	60.9	1.6	38.1
LG	4.6	(1.4)	38.5	0.6	43.7	10.5	4.2
Lotte	3.6	(1.3)	75.0	0.5	79.2	4.5	17.6

Note 1. *Chongsu* family = same person + relative/spouse; figures in parentheses indicate other relatives
2. Others = executives + non-profit corporations + treasury stock
3. Shares = common stock + preferred stock
Source Korea Fair Trade Commission (2022) and author's own creation

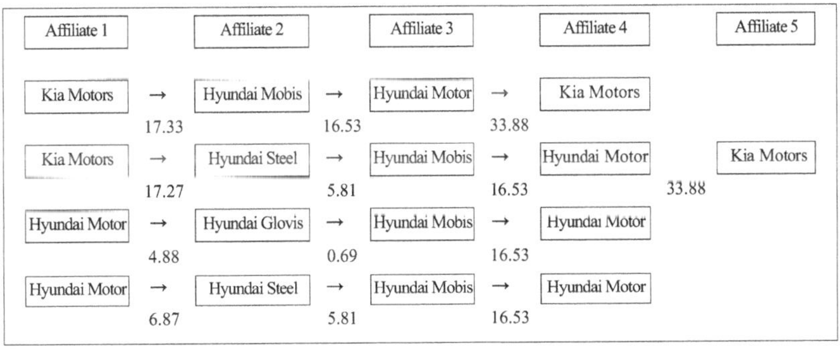

Fig. 2.5 Roundabout shareholding structure with multi-cores (%): Hyundai Motor Group (as of the end of 2021) (*Note* Shares = common stock + preferred stock. *Source* Author's own adaptation from Korea Fair Trade Commission, 2021)

to issues of economic justice and value conflicts so-called *chongsu*-holder value and control-holder value.

The values of the *chongsu* and control-holders are entrepreneurial and developmental with a view to promoting growth, but can also be irrational at times in order to maintain the identity of the *chaebol* system as a blood-related and diversified group. In many cases, there is a belief that the management rights should be passed on to the next generation of the head family. When this occurs, losses due to private defrauding or business failure spread throughout the group and even the entire business ecosystem through extended quasi-vertical integration, while profits are reaped by the *chaebol* families themselves, resulting in the appropriation of profits and socialization of losses.

In the *chaebol* governance structure, there is almost no difference in interests between the *chongsu*'s family and executives or managers, so it is possible to make quick and bold management decisions due to the low agency cost between them. However, the relationship between the executive and labor is hierarchical and vertical, with the *chongsu* at the top of the pyramid. In addition, a complex cross-shareholding structure centered around several core companies forms a kind of portfolio and lowers risk, but also creates the high systemic risk that a problem at one company could spread to the entire group. This cross-shareholding structure also enables external control through affiliates. In other words, the *chaebols* utilize the quasi-internal market to blur the legal boundary of the firm, i.e. to exploit organizational arrangements in which necessary resources are mobilized and distributed across the business group just-in-time and place in a way that transaction costs are reduced and organizational flexibility is improved to ease the rigidity of the group-wide network. Moreover, *chaebols* can always pass risks and costs on to external companies through the asymmetric exercise of economic power due to their monopsonist position on demand in the market.

As described above, subsequent quasi-vertical integration has formed through a complex multiple and pluralistic roundabout cross-shareholding structure between affiliates. Hyundai Mobis, a key affiliate of the Hyundai Motor Group, exclusively supplies key parts to Hyundai Motor or Kia Motors. By delivering modular parts produced by multiple external suppliers, Hyundai Mobis appropriates profits within the group and limits the bargaining power of external suppliers (see Fig. 2.6). Although the group's affiliates are legally independent companies, they have enjoyed the economic advantages of the *chaebol* through quasi-internalized links. In this way, Hyundai Motor has seized business opportunities resulting from modularization and fully controlled and dominated sub-parts suppliers.

Pre-Chandlerian *chaebols*, in which ownership and management are not separated, internalize the inertia of Chandlerian corporate organization in that they utilize resources within the *chaebol* through a closed structure that clearly distinguishes between the inside and outside of the organization. However, in some cases, they rely on outsourcing in a strategic and maneuverable manner. Since this structure could entail considerable transaction and adjustment costs in creating a social division of labor, where there is the need to invest in a specific asset, *chaebols* would prefer a strategy of mobilizing internal resources or *de facto* internalizing the external economy into the business group to that of using external inter-firm relationships. As shown in Fig. 2.6, subsidiaries in the producer service sector, such as finance and knowledge-based services, are in charge of the group's overall market demand. In this way, *chaebols* reduce costs and accumulate technology across the group by *de facto* internalizing key materials and parts from affiliates through quasi-vertical integration. In addition, this is sometimes tacitly aimed at solidifying the succession structure of the group's weak corporate governance or promoting private defrauding through the rapid growth of new affiliates in a short period of time via insider trading.

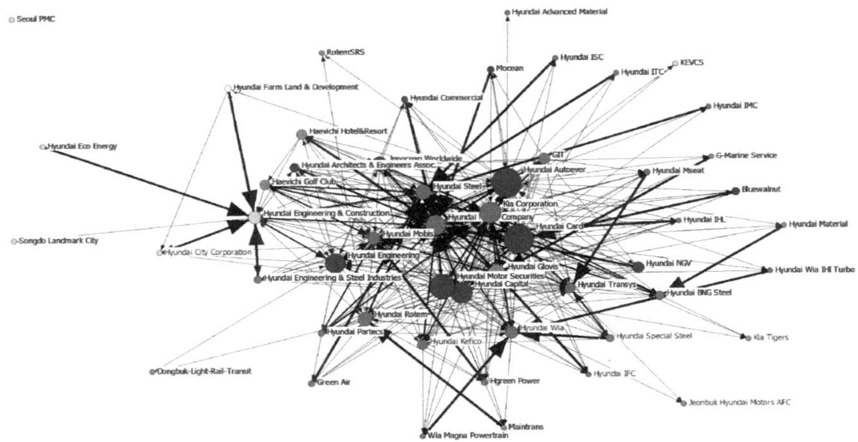

Fig. 2.6 Quasi-internal transactions of Hyundai Motor Group: Hyundai Motor's claim-obligation relationships with affiliates in relative terms (as of balance in 2021) (*Note* The arrow indicates the direction of the claim-obligation relationship, and its thickness represents the share of the remaining company's debt between affiliates. The color of the circle indicates the industry classification of the company, and its size represents the level of external connectivity. *Source* https://dart.fss.or.kr/ and author's own creation)

Since the 1997 Asian Financial Crisis, management transparency has been improved through various reforms to the corporate governance of *chaebol* conglomerate. Despite concerns that such reforms could lead to a 'crisis of governance amid excess governance,' which refers to the possibility of hostile mergers and acquisitions against *chaebol* conglomerates due to the massive influx of foreign capital, it appears that foreign investors and the *chaebol* system coexist with one another in a system called 'productivist neoliberalism' (Lee, 2013). This reflects the fact that *chaebols* have improved their management performance and adopted shareholder capitalism management behavior since the financial crisis, including increasing dividend payments and repurchasing treasury stock, with being aware of checks by external shareholders.[5]

2.2.4 Product Strategy and Profitability—The Pursuit of Flexible Scale-Diversity

The foundation of a production system is ultimately profitability. Boyer and Freyssenet (2000, 2002) list six revenue sources, namely scale, variety, quality, flexibility, innovation, and constant cost reductions, and present the concept of profit strategy as a combination of these. Among the profit strategies of top automobile makers, the Sloan model that combines scale and diversity (e.g. Volkswagen), the Honda model combines innovation and flexibility, and the Toyota model creates consistent cost reductions. The MIT-IMVP suggested that the profit strategies of all automakers would ultimately converge with Toyota's (Womack et al., 1990), but in reality, hybrid models exist through a combination of various revenue sources.

Profitability is not a simple quantitative indicator of corporate performance but needs to be understood as the result of compromises between different strategies implemented by a company and the various actors surrounding it. Certain factors cancel each other out. For a production system to be sustainable, a certain level of profit must be obtained. From an accounting point of view, profitability can be defined in several ways for various purposes. For example, markup (gross profit margin ratio) measures the company's pure operating performance, while return on equity (ROE) gauges the efficiency of asset use. The latter is a key

[5] Hyundai Motor received the 'Global Automotive Shareholder Value Award' from Automotive News in 2003, 2005, and 2011.

measure of shareholder capitalism, serving as an indicator of the short-term claim of return on capital (shareholder equity), regardless of how the business is run, rather than its potential for long-term growth (Aglietta & Berribi, 2007).[6]

Hyundai Motor's ROE was trending downward prior to the 1997 Asian Financial Crisis but bounced back in the 2000s except for the Global Financial Crisis in 2008, then began to decline once again in the 2010s. In contrast, the company's markup moved sideways in the 1980s but rose throughout the period from the 1990s until the mid-2010s, except for the mid-2000s (see Fig. 2.7). This is largely attributable to the expansion of the Korean market, the launch of Hyundai Motor's own model, and maintaining a monopoly position in the domestic market. This suggests that Hyundai Motor's market flexibility has improved over time.

However, the drop in markup since the mid-2010s reflects Hyundai Motor's difficulties in the US and Chinese markets and the limitations of its strategy as a fast follower. In the mid-to-late 2010s, the company failed to read the changes in the US market, which was shifting toward SUVs, and expanded its marketing strategy to focus on sedans, a category that saw sales volume plummet. In China, Hyundai Motor's sales volume also nosedived in the aftermath of the THAAD fiasco connected to geopolitical conflict between the US and China, and the figures have not recovered since that point as the company has failed to release differentiated products that can compete with other automakers.

In the 2000s, Hyundai Motor succeeded in achieving rapid quantitative growth through quasi-vertical integration of powertrains through the affiliates Hyundai Powertech, Hyundai Dymos, and Hyundai Wia, as well as chassis parts through Hyundai Mobis. In particular, it has acquired competitiveness through internalization of powertrains which previously had have a high defect rate.

Another factor that contributed to declining profitability in the 2010s was the heavy burden of labor costs resulting from a court ruling requiring 10,000 or so in-house subcontract workers to be given full-employment benefits (Jo & Jeong, 2021). Due to these factors, Hyundai Motor's ROE has fallen since the early 2010s, and markup has also declined with a time lag. However, a turnaround in profitability has been expected since 2019 thanks to the successful launch of the luxury model

[6] It is said that Warren Buffett prefers companies with a steady ROE of 15% or higher when investing in stocks.

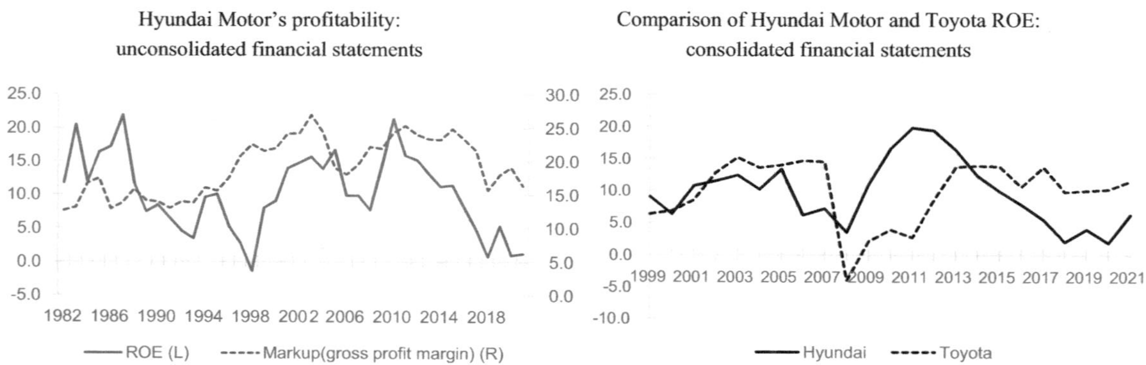

Fig. 2.7 Hyundai Motor's profitability (%) (*Sources* KIS-VALUE [http://www.kisvalue.com], Toyota's financial statements [https://global.toyota/en/ir/financial-results/archives/], and author's own creation)

Genesis in the United States, an enhanced brand image due to improved quality, and brisk sales of electric vehicles.

Toyota's ROE remains stably between 10 and 15%, excluding the recall period over 'pedal gate' from 2009 to 2010. Hyundai Motor made rapid progress during the Toyota crisis (see Fig. 2.7), although the former's ROE is highly volatile. As shown in Fig. 2.8, Hyundai Motor Group's share of global automobile production declined prior to the 1997 Asian Financial Crisis before subsequently rising, and the gap with Toyota narrowed the most in the late 2000s and early 2010s. The company's relative stock price also exceeded that of Toyota during this period. Hyundai Motor was also listed on Boston Consulting Group (BCG)'s list of the '50 Most Innovative Companies' during this period and was recently back on the list as its profitability improved (see Fig. 2.9). This shows that Hyundai Motor's second catch-up strategy in the 2000s succeeded and reached a peak in the early 2010s. However, unlike Toyota, Hyundai Motor has not been able to consistently and stably maintain this performance.

Hyundai Motor's markup is closely related to exchange rate fluctuations (see Fig. 2.10), with a correlation of −0.557 between 1982 and 2021, −0.558 between 1982 and 1999, and −0.547 thereafter, indicating Hyundai Motor's competitive strategy is sensitive to price fluctuations. This is mainly due to the production and sale of mid-priced products and also suggests that sustainable profits are only obtainable when economies of scale are pursued. After the merger between Hyundai Motor and Kia, the companies carried out platform integration work to reduce development costs and achieve economies of scale (see Table 2.2). Hyundai Motor went from 22 platforms in 2002 to just six integrated platforms in 2015. This implies that similar to Volkswagen, since the 2000s Hyundai Motor and Kia have shifted toward a volume and diversification strategy that reduces the number of platforms while increasing the number of models.

While pursuing economies of scale through platform integration, Hyundai Motor has simultaneously expanded its product lineup through ongoing technology development and building up the technological capabilities to produce a wide range of vehicles. As shown in Table 2.3, the number of Hyundai Motor–Kia products has increased significantly since the 1990s. This implies the increasing ability to flexibly respond to market demand. Figure 2.11 shows that compared to Toyota, Hyundai Motor has a high proportion of small and medium-sized cars and a lower proportion of recreational vehicles (RVs), but has various product lines such as

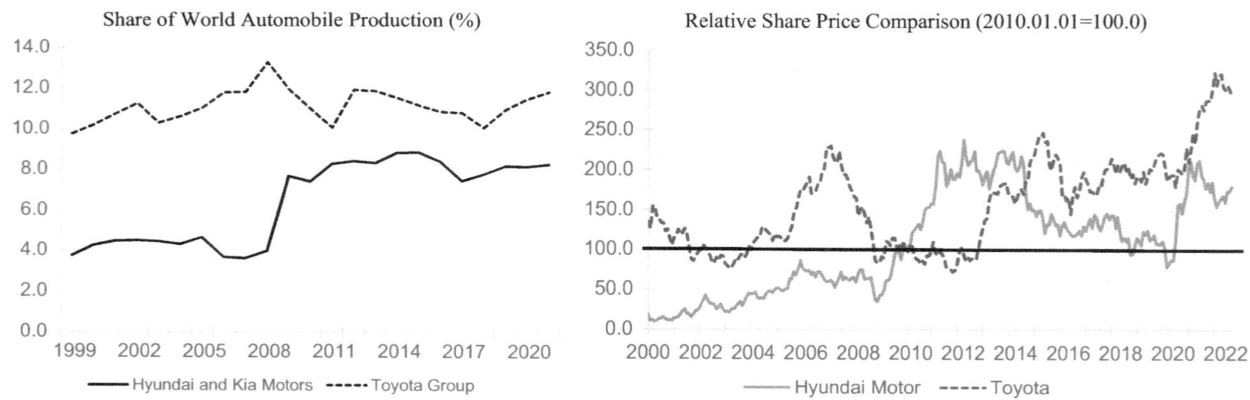

Fig. 2.8 Comparison of production and relative stock price between Hyundai Motor and Toyota (*Note* Production share is OICA [Organisation Internationale des Constructeurs d'Automobiles] data until 2018, KAMA [Korea Automobile Manufacturers Association] data from that point on. For the latter case, the 'world as a whole' is the aggregation of the United States, Canada, United Kingdom, Germany, Spain, France, Italy, Brazil, Japan, China, India, Russia, Mexico, and Korea. *Source* OICA [https://www.oica.net/category/production-statistics/], KAMA [https://www.kama.or.kr/], Yahoo Finance [https://finance.yahoo.com/], and author's own creation)

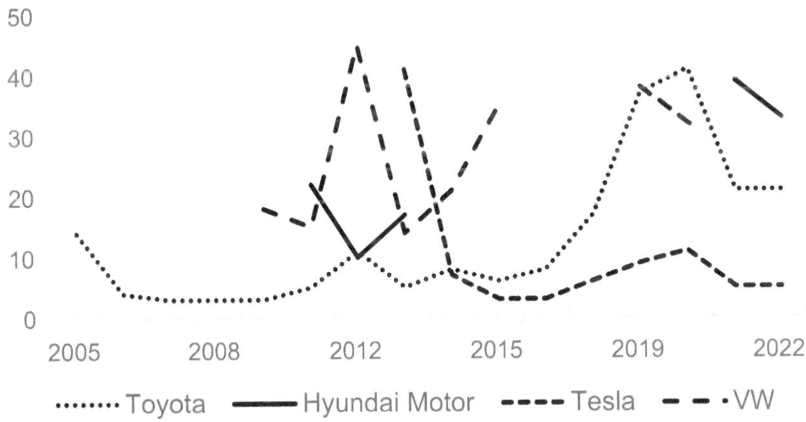

Fig. 2.9 Comparison of several automakers and auto parts companies among BCG's 50 most innovative companies (*Source* BCG [https://www.bcg.com/publications/most-innovative-companies-historical-rankings] and author's own creation)

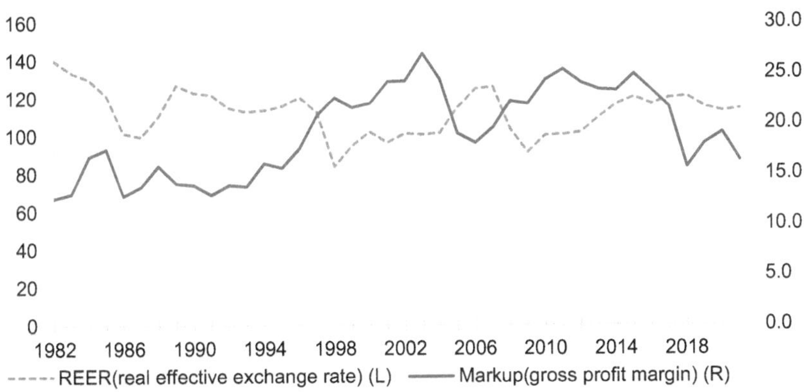

Fig. 2.10 Correlation between Hyundai Motor's markup and exchange rate (*Note* Based on unconsolidated financial statement. *Sources* KIS-VALUE [http://www.kisvalue.com], BIS [Bank for International Settlements] [https://www.bis.org/statistics/eer.htm], and author's own creation)

Table 2.2 Hyundai Motor-Kia Motors platform integration

	2002	*2009*	*2011*	*2015*
Integrated platforms	0	6	6	6
Total platforms	22	18	11	6
Total vehicle models	28	32	36	48

Source Hyundai Motor's IR data in 2019 and author's own creation

low-end vehicles, high-end vehicles, and RVs. The geographic portfolio of products is also balanced. However, although RV models have become mainstream in the US market since the mid-2010s, the limitations of the product lineup have led to a fall in profitability. Nevertheless, the company still aims for a turnaround in profitability by improving its RVs and electric vehicles (EVs) product lineups.

Since the 1990s, Hyundai Motor has expanded its product lineup based on a monopolistic position in the Korean market and self-developed models, continuing to raise markup until the mid-2010s. This profit strategy of flexible scale and diversity is linked to quality improvement across all products. Competitiveness in the automobile industry can be divided into price and quality, and Hyundai Motor's main problem was competing on quality. Figure 2.12 shows the trends in J.D. Power's relative Initial Quality Study (IQS) and Vehicle Dependability Study (VDS) for Hyundai Motor, compared with the industry average. Both indicators reveal that Hyundai Motor's quality has improved significantly, to the point where it is overtaking Toyota.

Table 2.3 Number of new products (passenger vehicles) by Hyundai Motor and Kia Motors

	Hyundai Motor	*Kia Motors*	*Total*
Prior to 1970	3	0	3
1970–1979	4	5	9
1980–1989	7	3	10
1990–1999	19	17	36
2000–2009	20	23	43
2010–2019	28	30	58
2020–2022.Q1	9	6	15

Source KAMA (https://www.kama.or.kr/) and author's own creation

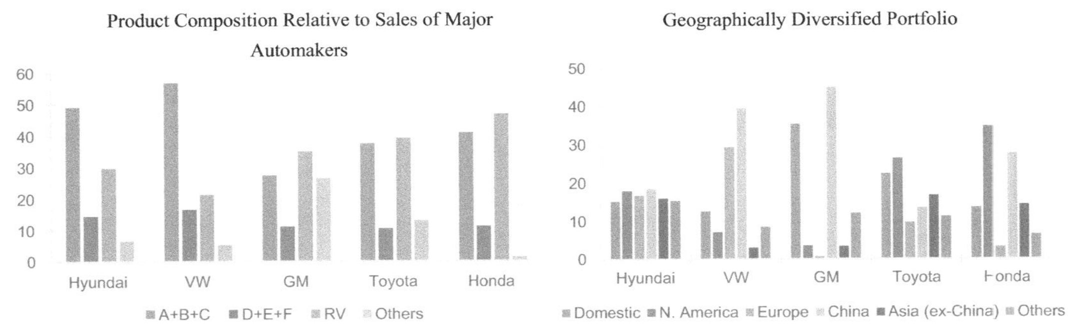

Fig. 2.11 Product composition of major automakers (%) (as of 2017) (*Note* A + B + C = Mini + Sub Compact + Compact; D + E + F = Mid-Size + Large + Luxury; RVs [Recreational vehicles] = SUVs [Sport Utility Vehicles] + MPVs [Multi-Purpose Vehicles]; Others = PUVs [Premium Unique Vehicles] and others; excluding HCVs [Heavy Commercial Vehicles]. *Source* Hyundai Motor's IR data in 2019 and author's own creation)

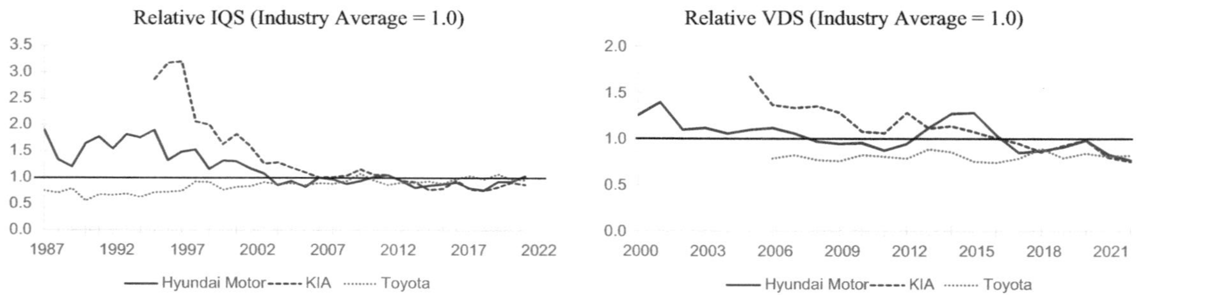

Fig. 2.12 Changes in relative quality index of Hyundai Motor and Toyota (*Source* J. D. Power [https://www.jdpower.com/] and author's own creation)

Since the late 2000s, both Hyundai Motor–Kia and Toyota have been below the industry average of 1.0 on the relative IQS index. In some years, Hyundai Motor is even ahead of Toyota. For the relative VDS, which gauges durability and quality, Hyundai Motor reached the level of industry average in the late 2000s, but in the first half of the 2010s, it returned the above industry average level again. A reversal has occurred for Hyundai Motor which stands below the industry average since the mid-2010s. In contrast, the relative VDS for Toyota has moved sideways without significant fluctuations since the mid-2000s. Note that since these indices indicate the number of defects, the lower the number, the better the quality. Thus, this shows that Hyundai Motor–Kia have seen rapid improvement in their ability to compete on durability and quality and that their rapid rise in the relative IQS and VDS rankings since the 2010s is unprecedented and highly unusual. On the other hand, Toyota has maintained stable competitiveness in durability and quality for a long time.

This improvement in quality was largely attributable to the *chaebol* head's quality-oriented management strategy that emerged in the late 1990s. In accordance with this policy, a dedicated team was established for each IQS item, and a structure was created in which testing, design, and manufacturing staff jointly identified the cause of problems and shared responsibility for fixing them. This is a different approach from Toyota, where producing zero defective vehicles through continuous improvement in the production line is the top priority, and quality is only the result of this constant improvement. On the other hand, improved quality at Hyundai Motor is the result of sequentially eliminating the causes of problems found by engineers. If Toyota is inductive, then Hyundai Motor seeks quality improvement in a deductive way, but both share the same goal. While Hyundai Motor's method is short-term and improvisational under management leadership, Toyota's is long-term, sequential, and cumulative through voluntary involvement from shop-floor workers.[7]

[7] Along with quality improvement, the brand image of Hyundai Motor-Kia has also rapidly risen. In Interbrand's 2021 global top 100 brand survey, Hyundai Motor and Kia ranked 35th and 86th overall, and 6th and 11th in the automotive sector category, respectively. In the automobile sector, companies higher than Hyundai Motor were Toyota, Mercedes-Benz, BMW, Tesla, and Honda (see https://interbrand.com/best-global-brands/).

2.3 A Micro Overview of Hyundai Motor's Production System After the 1997 Asian Financial Crisis

This section provides a brief overview of Hyundai Motor's production system, including engineer skills, work organization, labor–management relations, inter-firm relations, and overseas production. More details of these characteristics are discussed in Chapters 3–6.

2.3.1 Skill-Building in Engineers: Learning by Improvisation

Engineering refers to a series of activities that logically organize product development and manufacturing processes through the application of science and technology. This can be divided into the fields of design technology (creating products), manufacturing design technology (laying the groundwork in preparation for mass production of new cars), and process technology (managing the automobile manufacturing process and incremental innovation). Since the Asian Financial Crisis, Hyundai Motor has concentrated on product development and production efficiency through a technocratic strategy led by engineers.

However, as shown in Fig. 2.13, Korean engineers only accounted for 7.4% of total employment in the automotive industry in 2018, a lower figure than Germany (17.9% in 2018) or Japan (11.2% in 2015). In particular, the proportion of computer scientists is relatively low in the Korean automobile industry. This suggests that Korean automakers are weak in basic advanced research and have acquired more capabilities in manufacturing design and process technology.

What is the reason for the low proportion of engineers employed by Korean automakers despite adopting a technocratic engineering strategy? This can be explained in three dimensions (Jo & Jeong, 2021). First, compared to Germany and Japan, Korean companies develop and produce mid-priced products, and engineers work long hours. As a result, their productivity is comparatively high.[8] This is shown in Fig. 2.14. As of October 2020, the number of active patents for Hyundai Motor and Kia is second only to Toyota. Figure 2.15 shows the ratio of R&D

[8] For example, a former high executive at Hyundai Motor's Namyang R&D Center said in an interview in the mid-2010s: "Compared to Volkswagen engineers, Hyundai Motor engineers work twice as much for half the salary."

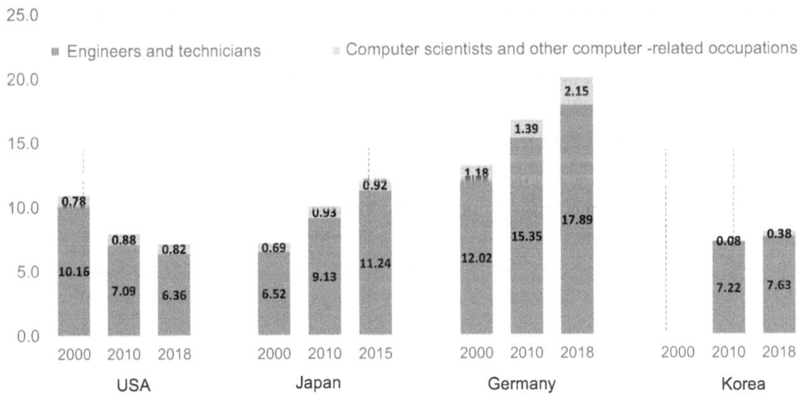

Fig. 2.13 International comparison of engineers in the automotive industry (%) (*Source* Author's own adaptation from Krzywdzinski [2020] and Jo and Jeong [2021])

investment to sales for both companies. Although time-series consistency is not guaranteed due to changes in accounting standards, the figure continues to rise over time except for the period immediately after the Asian Financial Crisis. However, there has been still a significant gap between Toyota and Hyundai Motor based on consolidated financial statements, which have not significantly changed since the mid-1990s. This suggests that Hyundai Motor's engineers are highly productive and cost-effective in that a small number of engineers have been able to produce a large number of active patents and design products and manufacturing processes.

Second, the number of engineers in Korean companies is underestimated. For example, in Hyundai Motor, engineers account for not only staff working in research institutes, product planning, manufacturing design technology, quality, and the strategic technology headquarters[9] but also the majority of the labor force in the Urban Air Mobility (UAM) division. Even though managers working in administrative departments are engineers, their job identity as engineers is not clear, so the higher the

[9] This was a newly established department at Hyundai Motor in 2016, and is in charge of investment feasibility and open innovation. The majority of staff are from the Research and Technology Center.

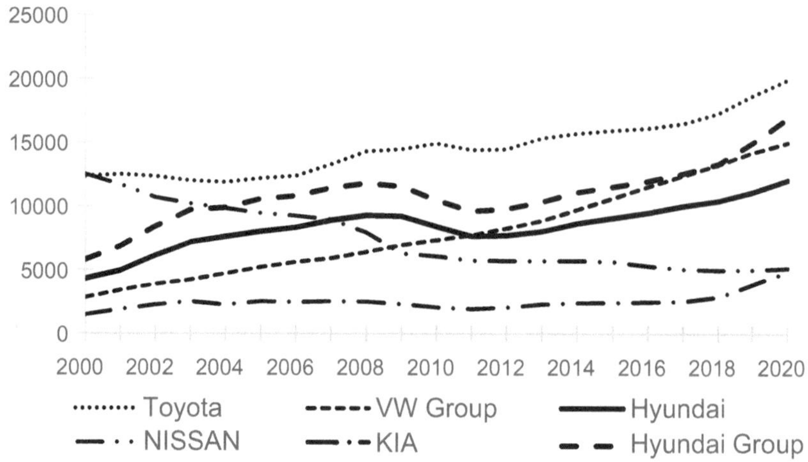

Fig. 2.14 Comparison of the number of active patents of major automakers (*Note* 1. The year of 2020 is as of October 2020. 2. Hyundai Motor Group = Hyundai Motor + Kia Motors. *Source* https://www.statista.com/statistics/1178549/number-of-patents-owned-by-the-top-automobile-manufacturers/, and author's own creation)

position, the more likely they would identify as administrative managers in employment surveys. This stems from the practice of HR management in Korean companies where engineers lack a strong job identity and do not have a separate promotions system.

Finally, labor market segmentation has been implemented within the engineering profession in the Korean automobile industry. As engineering work at automakers is outsourced to the group's affiliates or external parts makers for the purposes of cost reductions or employment flexibility, the ratio of engineers at automakers is underestimated. For example, engineers belonging to affiliates of Hyundai Motor Group, such as Hyundai Mobis, Hyundai Wia, and Hyundai Transys, are effectively in charge of engineering work at Hyundai Motor. Similar to the United States, Korea also has a high rate of engineering outsourcing. The low ratio of engineers in the United States is due to the high degree of outsourcing in product development engineering.

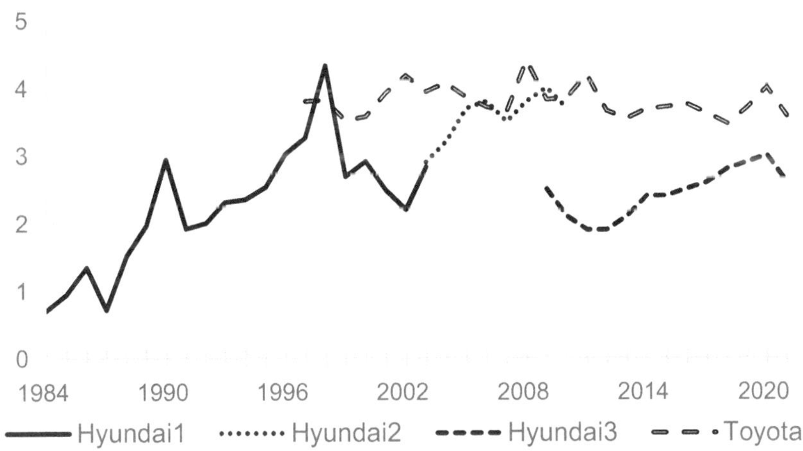

Fig. 2.15 Comparison of R&D investment to sales ratio between Hyundai Motor and Toyota (%) (*Note* 1. For Hyundai Motor, it should be noted that there is a disconnection in the time series due to changes in accounting standards: 1984–2003 is based on unconsolidated financial statements (Hyundai1), while 2004–2021 is based on consolidated financial statements (Hyundai2), and figures since 2009 have been disconnected from previous figures due to changes in accounting standards on a consolidated basis (Hyundai3). 2. For Toyota, figures are based on consolidated financial statements. *Source* KIS-VALUE [http://www.kisvalue.com], Hyundai Motor Annual Reports, Toyota's financial statements from [https://global.toyota/en/ir/financial-results/archives/], and author's own creation)

As Amsden (1989) notes, investment in engineer education and training through the formation of a managerial hierarchy played a decisive role in the growth of Hyundai Motor. Since the investment-led economy requires huge fixed capital investments and the latest manufacturing process design to handle them, the more emphasis is placed on the governance structure of big firms that can accommodate financing of such investments and the commitment of manufacturing design and process-centered engineers rather than skills in the workplace (Levy & Kuo, 1991). If this is the case, what are the conditions that lead to the formation of engineer-led organizational capabilities? First, Hyundai Motor acquired Kia and become a *de facto* monopoly in the domestic market after the Asian Financial Crisis, making it difficult for engineers to find a better job outside of Hyundai Motor. Second, the voluntary and sometimes involuntary commitments of engineers have been promoted by an internal promotions system, as well as restrictions on their inter-firm mobility due to relative high-wage levels in Hyundai Motor compared with those of competitors in the automotive industry in Korea. Finally, there has no association representing the interests of engineers, and many prefer being promoted to higher-ranked administrative managers instead of maintaining an identity of engineers (Kim et al., 2022). Using Hirschman's (1970) famous framework of 'exit, voice, and loyalty', it can be explained that engineers in monopolistic firms may have limited opportunities to leave for other firms in the same sector (exit) and may lack representation for their interests within the organization (voice). As a result, they may choose to pursue promotions to managerial positions and voluntarily dedicate themselves to the company's growth and success (loyalty).

In this context, committed engineers build organizational capabilities by developing their skills within Hyundai Motor. The divide between technology and skills that Hattori (1986) noted in the 1980s has continued since then, with a divergence in skill formation between engineers and shop-floor labor. This means that some engineers and skilled workers fill the void in skills among ordinary shop-floor workers, resulting in an excessive workload for engineers (Cho, 2014).

'Learning by improvisation' has been a major part of skill formation among engineers at Hyundai Motor. The integration of freedom (flexibility) and control (efficiency) has been considered a major challenge for product innovation in today's automotive firms (Clark & Fujimoto, 1991). Improvisation refers to the organizational capability to act spontaneously in an attempt to respond to problems in an innovative way through a combination of organizational autonomy (freedom) and goal clarity (control) in a turbulent environment.

The concept of improvisation contains four dimensions. First, improvisation is associated with the minimal structure in macro routines, meaning that a clear goal is combined with autonomy. Second, improvisation involves the recombination of existing available resources such as experience and expertise, indicating an ability to combine resources in unconventional ways and execute actions within time constraints. Third, it involves a rapid response, referring to the ability to respond to changes in

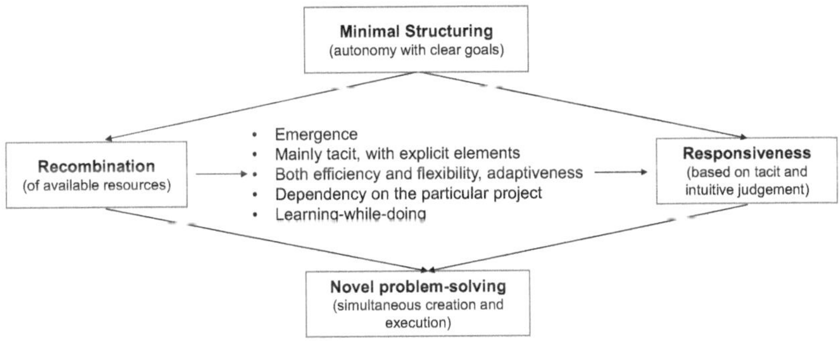

Fig. 2.16 Dimensions of organizational improvisation (*Source* Author's own adaptation from Kim et al., 2022)

the environment in a timely manner. Finally, improvisation entails innovative solutions where new ideas or processes are relevantly created to solve problems in specific situations (see Fig. 2.16).

The formation of organizational capabilities in Hyundai Motor was not based on the accumulation of routines in an orderly and sequential manner, but rather through repeated improvisation (Gong et al., 2006). Hyundai Motor acquired these novel capabilities through attempts to imitate and improve on foreign imported technologies.

Although Hyundai Motor's top management has focused on the abstract goal of producing high-quality vehicles under the 'quality management' slogan since the 1997 Asian Financial Crisis, the company has built dynamic organizational capabilities by quickly solving problems through trial and error, improvisational learning, and cooperation and deliberation with other stakeholders. This demonstrates one aspect of Herrigel (2020)'s experimentalism, where problems are solved through the mutual participation of related stakeholders. How is this possible in a business organization with a vertical hierarchy such as a *chaebol*? One reason is that Hyundai Motor's top management does not know and understand the specific processes of new car development and mass production in detail. Unlike leaders such as the late former Volkswagen chairman Ferdinand Piëch, many are not engineers with a strong background in car design.

Hyundai Motor has adopted an implicit organizational practice of delegating a certain amount of authority to engineers, without paying much attention to the mistakes that occurred in this process. As a result, engineers on the shop-floor have a voice on par with that of financial managers. This practice provides a space for autonomy to encourage experimental solutions within the organization, and, as will be discussed in detail in Chapter 3, the construction of the pilot center has been important in remarkably upgrading engineers' skills. The pilot center has also become a sort of reflective and meeting place where various stakeholders such as even external parts suppliers have easily brought together, thanks to the *chaebol*'s unspoken market power, to discuss the development and mass production of new cars through a deliberation process. This collaborative approach aims to quickly solve problems.

2.3.2 Work Organization and Shop-Floor Workers: Flexible Automation, Skill-Saving, and Segmented Labor

For latecomers such as Hyundai Motor, it was not easy to acquire the design capacity to quickly develop various models and foster the necessary shop-floor skills. Accordingly, the company pursued a production system that differs from Toyota. Since the 1990s, the ambitious goal set by Hyundai Motor's engineers to catch up with advanced companies has been none other than the establishment of a flexible production system. The company-wide adoption of computer-aided design (CAD), computer-aided manufacturing (CAM), and computer-aided engineering (CAE) helped to further this goal, as well as high-precision materials requirements planning (MRP) and bill of materials (BOM). In the 2000s, Hyundai Motor completed a flexible production system capable of producing 2–3 models at the same time on each production line in all plants through general-purpose equipment and software (Lee & Jo, 2007). The real value of this system has come to the fore as Hyundai Motor's ratio of exports increases, and the rapid growth of the company's plants suggests that the efforts of engineers have been successful.

Behind Hyundai Motor's emphasis on automation lies a production philosophy in which production is not overly dependent on workers. Tasks that require human labor are divided and simplified as much as possible (Oh, 2022).[10] This production system is powerful in the early

[10] GM and Fiat attempted to automate their final assembly lines to cope with productivity and quality problems in the 1970–1980s. At that time, labor–management relations in these companies were not smooth and it was not easy to maintain control over the

growth stage but can lead to problems down the line in mature markets or when market demand fluctuates. Since a fixed capital investment replaces labor and mobilizes large amounts of capital, it naturally requires long working hours and continued increases in sales volume to heighten the capacity utilization rate and subsequently recoup this investment. In other words, external sales growth is needed for Hyundai Motor to pursue economies of scale based on automation[11] and process simplification. For example, profitability fell sharply in the mid-2010s when sales did not increase as expected in the Chinese and US markets.

Figure 2.17 shows Hyundai Motor's fixed capital investment against cash flow. Before the Asian Financial Crisis, the company adopted a high-debt, high-investment model irrespective of cash flow, but fixed capital investment has been made in a way that retains greater cash flow than an investment since then. This fluctuates according to Hyundai Motor's growth pattern. Although the propensity to invest as the ratio of fixed capital investment to operating profit had generally fallen by the 2010s even with some slight reversals, it has been on the rise since then (see Fig. 2.18). Capacity investment rebounded in the 2010s due to investment arising in response to rising labor costs, as well as investment in new business opportunities including electric vehicles, hydrogen vehicles, and mobility transitions.

In addition, robot density defined as the number of operating industrial robots per 10,000 employees in the Korean automobile industry has increased sharply during the 2010s. This results not only from rapid wage increases but also confrontational labor relations and a sharp fall in the price per unit of robots due to technological advancement and greater demand. As a result, jobs including welding, loading, and inspections have

workplace. This automation was directed toward the engineering ideal of an unmanned factory (Camuffo & Volpato, 1997). Volkswagen's automation was an engineering quest to increase efficiency through great advances (Jürgens et al., 1997). Toyota, which had high productivity and quality due to continuous improvements to the production process, had the lowest level of automation at the time (Fujimoto, 1997). Since then, Toyota has exploited a human fitting automation strategy in response to worker resistance to the high-intensity labor process inherent to the lean production system. In contrast, Hyundai Motor has invested in capital-intensive automation that leverages engineers to increase production and improve quality.

[11] Automation investment is actively supported by labor unions due to the tedious and laborious nature of work processes, and IT-based automation acts as a mechanism to standardize quality to offset the lack of shop-floor skills.

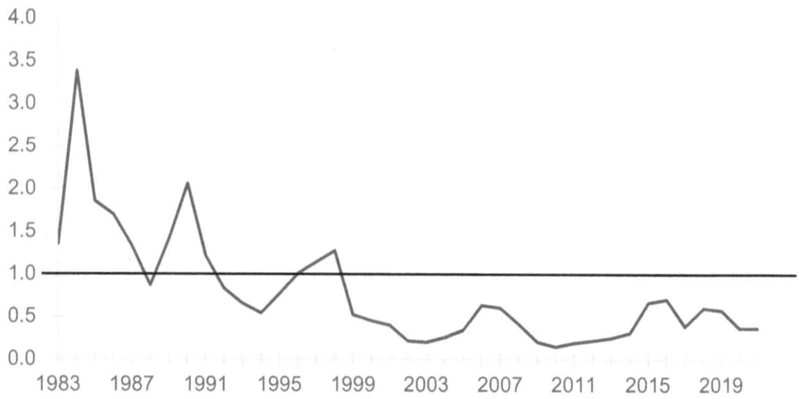

Fig. 2.17 Hyundai Motor's fixed capital investment compared to cash flow (*Note* The figure is defined as the ratio of fixed capital investment to cash flow using unconsolidated financial statements. *Source* KIS-VALUE [https://www.kisvalue.com/] and author's own creation)

rapidly been replaced by robots (Jo & Jeong, 2021). These trends suggest that automation is still a key element in Hyundai Motor's production system.

In a mass production system such as Fordism, a push system controls a process flow in which one process supplies parts to the next process. In contrast, the Toyota production system uses a pull system, where the subsequent process takes parts from the previous process. In the latter, since tasks are defined beyond the moving assembly line within the plant, i.e. through inter-firm relationships outside the plant, the flow of manufacturing processes is totally synchronized across the inter-firm boundaries. Toyota pulls as much as is needed from back processes to front processes to produce 'what is needed when it is needed,' also known as just-in-time, while Hyundai Motor pushes from front processes to back processes as seen in Fordism (Lee & Jo, 2007).

One of the obstacles that prevented Hyundai Motor from fully absorbing the Toyota production system before the Asian Financial Crisis was the lack of institutional complementarity between the work organization and the employment system. In Toyota's production system, full participation from workers is needed to continually eliminate waste and

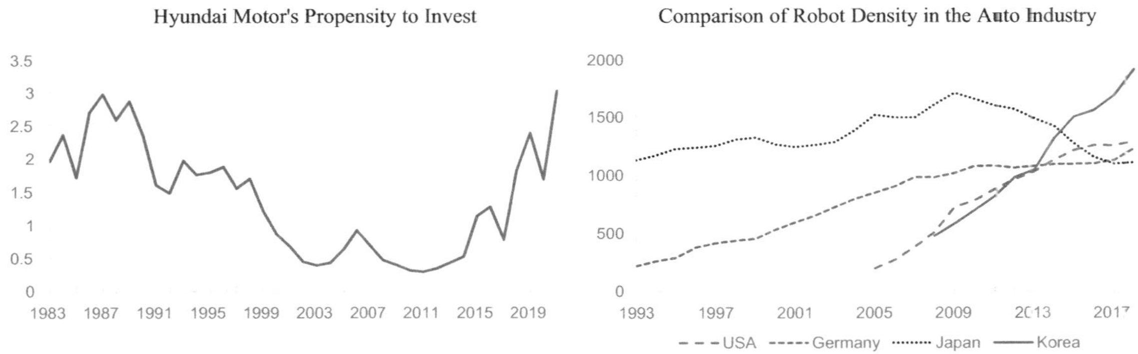

Fig. 2.18 Hyundai Motor's investment propensity and robot density in the automotive industry (*Note* Propensity to invest is the ratio of fixed capital investment to operating profit using unconsolidated financial statements. *Source* Author's creation and adaptation from KIS-VALUE [https://www.kisvalue.com/], Krzywdzinski [2020], and Jo and Jeong [2021])

improve processes, and long-term employment and cooperative labor–management relations are required to support this. As described above, labor relations have always been hostile at Hyundai Motor. Hyundai Motor's business strategy of flexible automation and information-based, engineer-centered rationalization mentioned above with regard to labor–management resulted in strengthening labor intensity through the application of the modular arrangement of pre-determined time standards (MODAPTS) method immediately after the employment adjustment in 1998. However, in the workplace where Tayloristic organizational configuration with a clear task divide between conception and execution is deeply embedded, the union has exerted great power over managers since the 2000s thanks to its strong organizing force, to some extent leading to the transformation of existing Taylorism firmly applied to the shop-floor. Hyundai Motor's management aimed to increase output and sales through huge amounts of flexible investment in automation. However, the union and its delegates gained relative power in the workplace due to their ability to threaten refusal to work overtime or engage in partial strikes. As a result, Hyundai Motor's management made a series of concessions to the union.

This power dynamic led to the creation of a work organization that differs from Toyota. For example, ordinary workers are mainly engaged in fragmented and simple repetitive tasks at Hyundai Motor. Although job rotation among shop-floor workers is carried out within groups every two weeks or once a month, this is used as a means of maintaining fairness or overcoming monotonous tasks rather than promoting skill development. Shop-floor workers with a long tenure have naturally come to embody firm-specific skills, but there has been a lack of effort to involve them in shop-floor improvements. Tasks such as improving productivity and quality are mainly carried out by upper-level workers such as keepers and foremen, as well as indirect workers in charge of tasks such as repairs and maintenance and quality control. Most ordinary shop-floor workers are not active participants (Jo & Paik, 2010).

As the Korean labor market became more flexible in the wake of neoliberal reforms after the Asian Financial Crisis, the labor union signed a full-employment protection agreement with the company in 2000 in which the utilization of non-regular workers was made available by the company if necessary, which refer to jobs that do not receive guarantees in terms of working method, period, and continuity of employment unlike regular workers in Korea. Hyundai Motor's regular or full-time worker-holder value has been obsessed with a sort of short-termism where employment stability and wage increases are on the priority list above all through the acceptance of the company's call for using non-regular workers and the sharing of the company's fruitful results through voluntary participation in long hours. Thus, instead of long-term skill development, a sort of hostilely commensal relationship called *synecthry* in the ecosystem has appeared between labor and management in which the fruits of the current period are shared.[12] The union of regular workers tolerates the harsh use of non-regular workers by the company and does not display much interest in the accumulation of skills in a broad sense.[13] As shown in Fig. 2.19, there was a positive correlation between Hyundai Motor's labor share and markup before the financial crisis, but it has been negative since then. This could be partially due to the trade union's pursuit of economic benefits such as wage increases based on its strong bargaining power.

Since the 1997 Asian Financial Crisis, Hyundai Motor has created a highly flexible production system by combining the latest technology designed by engineers with low-skilled workers. Although the company promotes IT-based flexible automation against the background of hostile labor–management relations, aggressive investment has not been carried out as it was in the past. Instead, flexibility and product diversity are acquired through the use of non-regular workers, regular or periodical demands for lower unit prices from parts suppliers, and modularization. However, in order for the company to recoup such fixed investments in automation as quickly as possible, there is a need to increase higher capacity utilization rates, and subsequently reach higher production levels without any interruption such as strikes. To this end, the company has no choice but to make a certain compromise with the strong union. This has become the material basis for the company's hostile coexistence with the

[12] Even though Hyundai Motor's workers are low-skilled, they have achieved wage increases and improved working conditions through the union's strong bargaining power, and paradoxically contributed to the establishment of a skill-saving production system by acquiescing to the company's increased employment of non-regular workers.

[13] Until the mid-2000s, Hyundai Motor's growth was centered around regular workers. Even though the tenure period for regular workers increased due to the pay scale, Hyundai Motor was able to maintain a constant ratio of non-regular workers and thus keep labor costs constant. Since the 2010s, the company has gradually converted 10,000 or so in-house subcontract workers into regular workers under court rulings, which contributed to deteriorating profitability (Jo & Jeong, 2021).

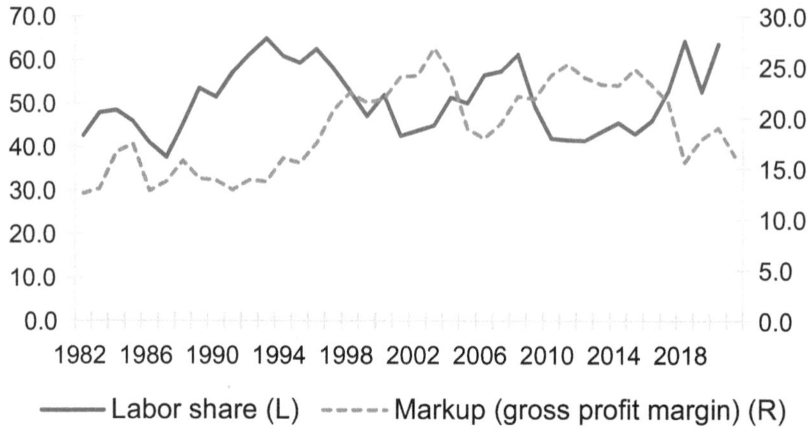

Fig. 2.19 Trends in Hyundai Motor's labor share (%) (*Note* 1. Based on unconsolidated financial statements. 2. Labor share indicates the ratio of wage to total added value composed of income and loss before income taxes, financial expense, tax and public dues, depreciation expense, wage, and rent. *Source* KIS-VALUE [https://www.kisvalue.com/], Korean Statistical Information Service [https://kosis.kr/index/index.do], and author's own creation)

union, creating a power configuration in the workplace where the union dominates. It is the very paradox of excessive automation investments which have attempted to replace and avoid labor. Since the 2000s, the union has also tacitly or explicitly agreed with the company's investment in automated facilities, with long working hours and the extensive use of non-regular workers in return for economic benefits.

2.3.3 Inter-Firm Relationships: Extended Quasi-Vertical Integration

Hyundai Motor is supplied with parts and materials through major affiliates that are part of a system we call quasi-vertical integration. Since Hyundai Motor enjoys a virtual monopsony on demand in the business ecosystem of the Korean automobile industry, it is possible for the company to form a system of relationships equivalent to those of affiliates with cooperative external suppliers. This means that *chaebol* conglomerates are able to exercise total domination and control over the parts and

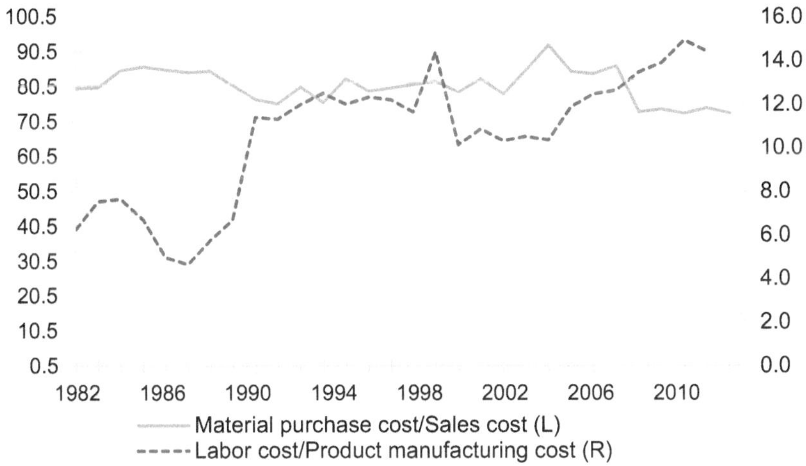

Fig. 2.20 Changes in Hyundai Motor's labor and material costs (%) (*Note* Based on unconsolidated financial statements. *Source* KIS-VALUE [https://www.kisvalue.com/] and author's own creation)

materials business ecosystem through a vertical and hierarchical transaction structure. For example, when the price of raw materials rises and there is greater uncertainty in market demand due to demand fragmentation or an economic crisis, *chaebols* have the bargaining power to pass this burden on to external suppliers through periodic unit price cuts to cope with such fluctuations. Figure 2.20 shows the relationship between Hyundai Motor's labor costs and material costs.[14] Since the 2000s, there is a clear intersection between the increase in labor cost burden within the company and the reduction in the cost of materials purchased from external parts makers. In other words, the company appears to have compensated for the increase in internal labor costs, which is difficult to control due to confrontational labor relations, through unit price cuts at parts makers.[15]

[14] Since manufacturing cost reports have not been released since the 2010s, it is difficult to ascertain the trend after that.

[15] In addition, various factors such as modularization at parts makers and lower raw material prices may have had an impact on this intersection.

Chaebols pass on market price fluctuations to external companies by taking advantage of their monopsony on demand, bargaining position as the final purchaser of parts, and multiple orders from core *chaebol* affiliates (see Fig. 2.21). For example, they not only pass on a portion of price fluctuations to external subcontractors and control their costs through cost-linked pricing but also siphon off profits from subcontractors and steal technological innovation. However, this extended quasi-vertical integration also helps to share and transfer risks. Since *chaebols* purchase parts and materials in a long-term relationship, they provide a stable market for partner companies and absorb market fluctuations even when market conditions are not good. Nevertheless, since the system is still closed in nature, it can distort competition by hindering fair trade or market opportunities for new entrants, causing the problem of highly concentrated economic power.

Modularization generally exploits external economies of scale and open networks (Sako, 2003; Sturgeon, 2002). *Chaebols* such as Hyundai Motor are able to solve the coordination problem of efficient allocation of resources by bringing external economies of scale into the business group through their affiliates and/or external partners. Even though these partners are legally external companies, they are effectively internal affiliates in

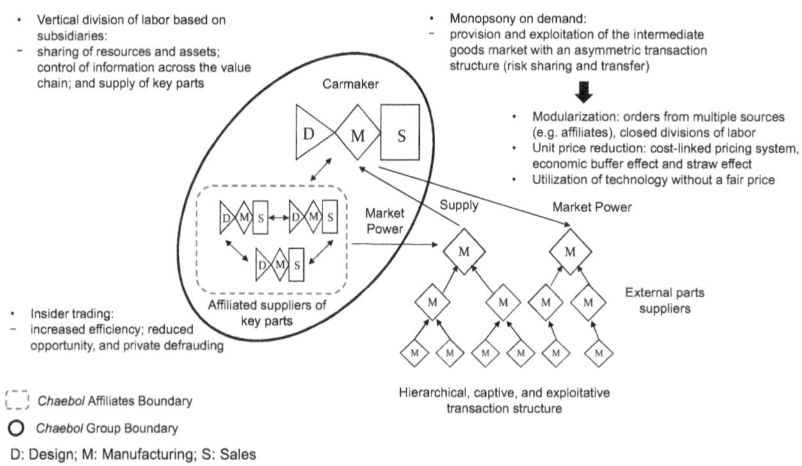

Fig. 2.21 Extended Quasi-Vertical Integration in the Korean automotive industry (*Source* Author's own adaptation from Jeong, 2022)

practice, so it is difficult to maintain the open nature of the network. Due to the nature of the Korean automobile industry, inter-firm networks are complex and multi-layered. In order to reduce transaction costs related to this and manage the network, *chaebols* may utilize *de facto* internal resources to absorb external shocks and ask affiliates and external partners to act as a buffer. Thus, there would be flexibility for *chaebols* to distribute the management burden of flagship companies in trouble across affiliates and build organizational capabilities to secure the group's core competencies through reorganization of subsidiaries and selection of outside partner companies by exercising asymmetric power. Although this is derived from the values of the *chongsu-holder* as the head of the company, it has quite uncertain experimental characteristics because this bold vertical decision-making by the head has real option value as the environment changes, as discussed in Chapter 1. Of course, there is a possibility of committing a Type 2 error at the same time. The former has a kind of rationality of efficient allocation of resources, but the latter does not.

2.3.4 *Overseas Transplant Strategy: The Geographic Transfer of Hyundai Motor's Production System*

Continued sales volume growth is required in order for Hyundai Motor's production system to enjoy both internal and external economies of scale. Accordingly, the company effectively had no choice but to engage in foreign direct investment. Hyundai Motor completed the construction of a local plant in Bromont, Canada in July 1989 with a view to use this as a stepping stone into the North American market, but this ultimately ended in failure. As a result, Hyundai had virtually no localized production before 1990, but this overseas production's proportion began to increase rapidly after the 2000s thanks to post-Cold War globalization. As shown in Fig. 2.22, Hyundai Motor–Kia's proportion of overseas production has slightly outpaced Toyota's since 2010.

Hyundai Motor's globalization strategy has been aggressive. In order to recoup the committed enormous fixed capital investment, Hyundai Motor's production system needs to create a positive feedback loop by selling more products in overseas markets beyond the domestic market. This was the rationale behind the announcement of 'quality management,' the large-scale recruitment of engineers to carry out this policy, and aggressive overseas expansion. Hyundai Motor preferred expanding

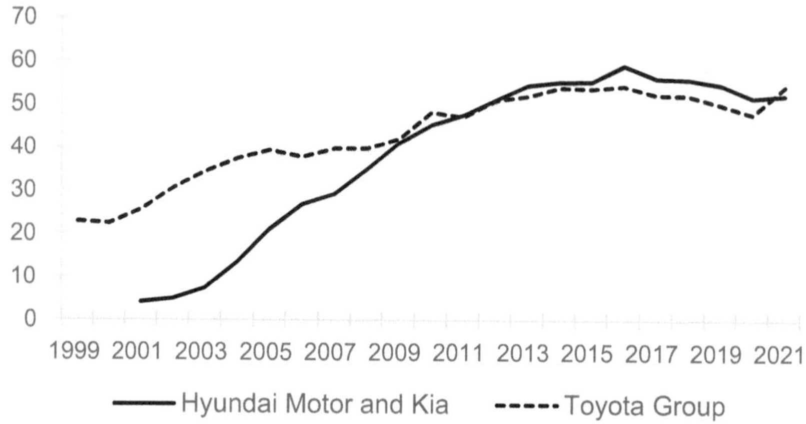

Fig. 2.22 Comparison of overseas production between Hyundai Motor Group and Toyota Group (%) (*Note* The Toyota Group includes Daihatsu and Hino. *Source* KAMA [https://www.kama.or.kr/], Toyota's financial statements [https://global.toyota/en/ir/financial-results/archives/], and author's own creation)

into green-field areas where it could start from scratch instead of leveraging existing local partners through M&As to lower risk. This green-field strategy centered on the construction of new overseas plants exploits low labor costs while also maximizing production flexibility. This is because green-field overseas plants in these marginal locations do not have to contend with local labor relations. Automakers in developed countries also take advantage of low labor costs in overseas plants, but they do not prioritize union-free management in the same way as Hyundai Motor. For example, Toyota's first US plant was acquired through a joint venture with GM called New United Motor Manufacturing Inc. (NUMMI), and GM's previous union was recognized by Toyota without any alterations (Jo & Jeong, 2021). This strategy of Hyundai Motor can be seen as a typical example of the 'spatial fix' that Harvey (1982) refers to.

References

Aghion, P., Guriev, S., & Jo, K. (2021). Chaebols and Firm Dynamics in Korea. *Economic Policy, 36*(108), 593–626.

Aglietta, M., & Berribi, L. (2007). *Desordres dans le Capitalisme Mondial*. Editions Odlie Jacob.
Amsden, A. (1989). *Asia's Next Giant: South Korea and Late Industrialization*. Oxford University Press.
Aoki, M. (2001). *Toward a Comparative Institutional Analysis*. MIT Press.
Aoki, M. (2010). *Corporations in Evolving Diversity: Cognition, Governance, and Institutions*. Oxford University Press.
Boyer, R., & Freyssenet, M. (2000). A New Approach of Productive Models: The World That Changed the Machine. *Industrielle Beziehungen [The German Journal of Industrial Relations], 7*(4), 385–412.
Boyer, R., & Freyssenet, M. (2002). *The Productive Models: The Conditions of Profitability*. Palgrave.
Camuffo, A., & Volpato, G. (1997). Building Capabilities in Assembly Automation: Fiat's Experiences from Robogate to the Melfi Plant. In K. Shimokawa, U. Jürgens, & T. Fujimoto (Eds.), *Transforming Assembly. Experiences in Automation and Work Organization* (pp. 167–87). Springer.
Chang, H.-J. (2004). Rethinking the Orientation of Economic 'Reform.' *Citizen and the World, 5*, 250–266. (In Korean).
Cho, S.-J. (2014). Success in Catch-up and Post Catch-Up Problem: Focusing on Hyundai Motor Group Case. *Journal of Korean Social Trend and Perspective, 91*, 136–168. (In Korean).
Christensen, C. M. (1997). *The Innovator's Dilemma: When New Technologies Cause Great Firms to Fail*. Harvard Business School Press.
Chung, M.-K. (1997). Introduction of New Production Methods and Changes in Human Resource Management: Focusing on the Case of Hyundai Motors Asan Plant. *Korean Journal of Labor Studies, 3*(1), 81–108. (In Korean).
Clark, K. B., & Fujimoto, T. (1991). *Product Development Performance: Strategy, Organization, and Management in the World Auto Industry*. Harvard Business School Press.
Cohen, W. M., & Levin, R. C. (1989). Empirical Studies of Innovation and Market Structure. In R. C. Schmalensee & R. Willig (Eds.), *Handbook of Industrial Organization* (pp. 1059–1107). Elsevier.
Fujimoto, T. (1997). Strategies for Assembly Automation in the Automobile Industry. In K. Shimokawa, U. Jürgens, & T. Fujimoto (Eds.), *Transforming Assembly: Experiences in Automation and Work Organization* (pp. 211–38). Springer.
Gerschenkron, A. (1962). *Economic Backwardness in Historical Perspective*. Harvard University Press.
Gong, Y. T., Baker, T., & Miner, A. S. (2006). *Capabilities and Routines in New Organizations: Evidence from the Field* (Working Paper. School of Business, University of Wisconsin—Madison).
Harvey, D. (1982). *The Limits to Capital*. Basil Blackwell.

Hattori, T. (1986). Technology Transfer and Management Systems. *The Developing Economies, 24*(4), 314–325.
Herrigel, G. (2020). Experimentalist Systems in Manufacturing Multinationals: Recursivity and Continuous Learning through Destabilization. In J. Glückler, G. Herrigel, & M. Handke (Eds.), *Knowledge for Governance: Knowledge and Space* (Vol. 15, pp. 415–439). Springer.
Hirschman, A. O. (1970). *Exit, Voice, and Loyalty: Responses to Decline in Firms, Organizations, and States*. Harvard University Press.
Humphrey, J. (2003). Globalization and Supply Chain Networks: The Auto Industry in Brazil and India. *Global Networks: A Journal of Transnational Affairs, 3*(2), 121–141.
Hyundai Motor Company. (1987). *20 Years of Hyundai Motor Company: 1967–1987*. Hyundai Motor Company. (In Korean).
Hyundai Motor Company. (1992). *A History of Hyundai Motor Company*. Hyundai Motor Company. (In Korean).
Jeong, J. H. (2022). Evaluation and Challenges of Korea's 'Industrial System' for Co-prosperity and Solidarity. In K.-Y. Shin & H. Yoon (Eds.), *Korea Trapped in Success: The Political-Economic Trajectory of Regressive and Selective Welfare* (pp. 69–138). Humanitas. (In Korean).
Jo, H. J. (1992). *A Study on the Production System of the Korean Automotive Industry: Focusing on Changes in the International Division of Labor in the 1980s*. Unpublished Doctoral Dissertation, Seoul National University. (In Korean).
Jo, H. J. (2005). *Is the Korean Production System Possible?: Exploring the Possibility of Hyundaism*. Hanul.
Jo, H. J., & Jeong, J. H. (2021). A Change in Employment Structure of the Korean Automobile Industry: 2008–2018. *Korean Journal of Labor Studies, 27*(1), 41–65. (In Korean).
Jo, H. J., & Paik, S. Y. (2010). The Change of Skill Character According to Flexible Automation of Production Process: Focused on the System Regulation Labor in Hyundai Motor Company's Body Shop. *Korean Journal of Labor Studies, 16*(1), 275–306. (In Korean).
Jung, S. G. (1995). The Politics of Flexible Production and Work Reorganization: The Case Study of Hyundai Motor Company. *Economy and Society, 28*, 219–245. (In Korean).
Jürgens, U. (2002). *Corporate Governance, Innovation, and Economic Performance: A Case Study on Volkswagen* (Working Paper, Social Science Research Center Berlin [WZB]).
Jürgens, U. (2003). Transformation and Interaction: Japanese, U.S., and German Production Models in the 1990s. In K. Yamamura & W. Streeck (Eds.), *The End of Diversity?: Prospects for German and Japanese Capitalism* (pp. 212–239). Cornell University Press.

Jürgens, U., Fujimoto, T., & Shimokawa, K. (1997). Conclusions and Outlook. In K. Shimokawa, U. Jürgens, & T. Fujimoto (Eds.), *Transforming Assembly. Experiences in Automation and Work Organization* (pp. 395–407). Springer.

Kang, M.-H. (1986). *Weird Koreans Who Made Pony*. Jungwoosa. (In Korean).

Kim, C., Jeong, J. H., & Jo, H. J. (2022). Improvisational Intensive Problem-Solving Capability: The Case of Hyundai Motor's New Car Projects. *Journal of Asian Sociology, 51*(4), 323–354.

Kim, G. (1994). *A Study on the Development of Technological Capability of Korea in the 1980s: With the Emphasis on the Development of 'In-house R&D System'* [Unpublished Doctoral Dissertation, Seoul National University]. (In Korean).

Kim, J. (2003). The Ownership and Governance of Korean *Cheabols*. *Journal of Korean Economic Development, 9*(2), 77–95. (In Korean).

Kim, L. (1997). *Imitation to Innovation: The Dynamics of Korea's Technological Learning*. Harvard Business School Press.

Kim, L. (1998). Crisis Construction and Organizational Learning: Capability Building in Catching-up at Hyundai Motor. *Organization Science, 9*(4), 506–521.

Kim, S.-J. (2006). Searching a New Coordination Mechanism at the End of Rapid Economic Growth, for Overcoming the Myth of the *Chaebol*-Centered Growth Model. *Economy and Society, 69*, 40–70. (In Korean).

Korea Fair Trade Commission. (2021, December 2). *Disclosing 'the 2021 Governance Structure of Public Announcement Targeting Business Groups'*. Press Releases. Fair Trade Commission.

Korea Fair Trade Commission. (2022, August 2). *Changes in Companies Affiliated with Large Corporate Groups for the Past Three Months*. Press Releases. Fair Trade Commission.

Krzywdzinski, M. (2020). *Automation, Digitalization, and Changes in Occupational Structures in the Automobile Industry in Germany, the United States, and Japan: A Brief History from the Early 1990s until 2018*. Paper presented to the GERPISA 2020 Conference.

Lee, B. C. (2013). Two-Pronged Structuration of Ownership, Control and Varieties of Capitalism: Based on the Critical Reformulation of the Theory of the Corporate Capitalism. *Journal of Korean Social Trend and Perspective, 89*, 222–258. (In Korean).

Lee, B.-H., & Jo, H.-J. (2007). The Mutation of the Toyota Production System: Adapting the TPS at Hyundai Motor Company. *International Journal of Production Research, 45*(16), 3665–3679.

Lee, K., & Lim, C. (2001). Technological Regimes, Catching-up and Leapfrogging: The Findings from Korean Industries. *Research Policy, 30*(3), 459–483.

Levy, B., & Kuo, W.-J. (1991). The Strategic Orientations of Firms and the Performance of Korea and Taiwan in Frontier Industries: Lessons from

Comparative Case Studies of Keyboard and Personal Computer Assembly. *World Development, 19*(4), 363–374.
Lung, Y. (2001). The Coordination of Competencies and Knowledge: A Critical Issue for Regional Automotive Systems. *International Journal of Automotive Technology and Management, 1*(1), 108–127.
Oh, J. (2022). *Research on the History of Technological Development of Hyundai Motor Company*. Meiji University. (In Korean).
O'Sullivan, M. (2002). Corporate Control. In W. Lazonick (Ed.), *The IEBM Handbook of Economics* (pp. 129–155). Thomson.
Ravenhill, J. (2003). From National Champions to Global Partners: Crisis, Globalization, and the Korean Auto Industry. In W. W. Keller & R. J. Samuels (Eds.), *Crisis and Innovation in Asian Technology* (pp. 108–136). Cambridge University Press.
Sako, M. (2003). Modularity and Outsourcing: The Nature of Co-evolution of Product Architecture and Organisation Architecture in the Global Automotive Industry. In A. Prencipe, A. Davies, & M. Hobday (Eds.), *The Business of Systems Integration* (pp. 229–253). Oxford University Press.
Sako, M., & Murray, F. (1999, October 6–7). *Modules in Design, Production and Use: Implications for the Global Automotive Industry* (Paper prepared for the IMVP Annual Forum, MIT Boston).
Song, H. K. (1999). *Labour Unions in the Republic of Korea: Challenge and Choice* [Discussion Papers (DP/107/1999)]. Labour and Society Programme, International Institute for Labour Studies.
Sturgeon, T., & Lester, R. (2004). The New Global Supply-Base: New Challenges for Local Suppliers in East Asia. In S. Yusuf, A. Altaf, & K. Nabeshima (Eds.), *Global Production Networking and Technological Change in East Asia* (pp. 35–88). Oxford University Press.
Sturgeon, T., Memedovic, O., van Biesebroeck, J., & Gereffi, G. (2009). Globalisation of the Automotive Industry: Main Features and Trends. *International Journal of Technological Learning Innovation and Development, 2*(1), 7–24.
Sturgeon, T. J. (2002). Modular Production Networks: A New American Model of Industrial Organization. *Industrial and Corporate Change, 11*(3), 451–496.
Sturgeon, T. J., van Biesebroeck, J., & Gereffi, G. (2008). Value Chains, Networks and Clusters: Reframing the Global Automotive Industry. *Journal of Economic Geography, 8*, 297–321.
Veblen, T. (1915). *Imperial Germany and the Industrial Revolution*. Macmillan.
Womack, J. P., Jones, D. T., & Roos, D. (1990). *The Machine that Changed the World*. HarperPerennial.
Yeo, I. (2021). Cooperation and Competition between the Korean and Japanese Automobile Industries. *Korean Journal of Japanese Studies, 13*(1), 52–83. (In Korean).

PART II

Empirical Anatomy

CHAPTER 3

Skill Formation in Engineers—Learning by Improvisation

3.1 Introduction

Amsden (1989) once referred to Korea as 'the next giant in Asia' in the sense that it is a country with the potential to successfully implement late-industrialization on a level comparable to Japan. It is true that in the early days of industrialization, the Korean economy had high expectations of a successful industrialization model by imitating Japan. However, the Korean economy has begun to take a different path from Japan since the late 1990s.

In this chapter, we explain how the Korean economy has taken a different path from Japan through the lens of the automobile industry, with a focus on the skills of engineers. We use the concept of skill to describe the distinctive capabilities of engineers.[1] Skill includes theoretical, empirical, and experiential elements. Engineers acquire knowledge in college and university, so theoretical elements are included in their skills. However, after joining an organization, they receive education and training and strengthen their own empirical and experiential skills in the company. Accordingly, in order to explain the skills of Hyundai Motor's

[1] If skill is understood in a broad sense to include not only job skills but also problem-solving and social capacities (McQuaid & Lindsay, 2005: 209–210), it is clear that not only ordinary workers but also engineers are bearers of skills.

H. J. Jo et al., *Agile Against Lean*,
https://doi.org/10.1007/978-981-99-2042-6_3

engineers, it is necessary to explain in detail how these theoretical, empirical, and experimental elements work together.[2]

Engineers demonstrate skills in several areas of the product value chain. Design information is applied to products through the production process and delivered to the customer. If we refer to generating design information as the product development process, mass production is the process of replicating this design in large quantities in a short period of time (Fujimoto, 2001: 28).[3] It is important to develop good products, but mass production that efficiently achieves good quality and productivity is equally important. It is vital for manufacturing design technology engineers to demonstrate their capability to efficiently produce developed products by organizing equipment and tools, parts, production methods, and workers. This is why the skills of manufacturing design technology engineers play a key role in determining the success or failure of a company.

Engineering skills are exercised through organizational capabilities at the collective level, not the individual level. Organizational capability refers to technical resources, human resources, and organizational routines as a common pattern of behavior that integrates and utilizes each type of resource efficiently (Nobeoka, 2007: 4). Accordingly, the organizational capabilities of a specific company can vary depending on how engineers collectively demonstrate their skills in organizational relationships with other actors. We seek to understand the collective skill of Hyundai Motor's engineers as an intensive problem-solving capability that includes timely, improvisational, and dynamic aspects.

Korean *chaebol* groups have developed their ability to execute projects by carrying out a variety of projects through diversification. Once acquired by the parent company, project execution capacity is spread to affiliates within the group through the dispatch of engineers, enabling affiliates to catch-up with competitors in more advanced countries. The

[2] For engineers, the theoretical element of skill refers to a series of steps used to solve problems in the field of science and technology. Professional education and training are needed to acquire these theoretical elements. The empirical element of skill refers to a series of steps for analyzing and solving problems based on field experience and observation rather than theoretical knowledge. In order to acquire empirical elements, hands-on experience is required through trial and error including observation, measuring, and experimentation.

[3] Products with integrated architecture such as automobiles need to finely coordinate design drawings and processes to achieve a certain level of quality and productivity.

Hyundai Group can be viewed as an exemplary success story of a late-comer adopting a strategy of diversification, as pointed out by Amsden and Hikino (1994). Project execution capacity manifests as the intensive problem-solving capabilities of engineers. The latter refers to the capability to solve problems by centrally mobilizing resources within a company to complete a specific project.

In the early days of industrialization, Hyundai Motor successfully acquired technological expertise that imitated the Japanese production system through Mitsubishi. Until the early 1990s, Hyundai Motor's new car development process employed a kind of horizontal manufacturing design technology engineering that was similar to that of Japan. In other words, the R&D center, manufacturing design technology engineers, process technology engineers, and/or technicians and shop-floor workers all worked closely together to develop new cars in preparation for mass production on an almost horizontal level.

However, since the late 1990s, Hyundai Motor has almost fallen into the 'middle-ranked maker's trap' as described in Chapter 2. Hyundai Motor encountered growth constraints with their current production system due to downturns in domestic and overseas conditions. In order to overcome this, Hyundai Motor changed course by replacing the company's CEO. At the heart of this move is an intention to promote innovation in the production system by putting the intensive problem-solving capabilities of engineers at the forefront of the company. The innovative solutions included the adoption of a modular production system and the construction of a pilot center within the site of the Namyang R&D Center. This chapter aims to show how engineers have been able to successfully demonstrate their collective skills since the construction of the pilot center.[4]

In this chapter, section two explains how Hyundai Motor decided to construct a pilot center in response to the middle-ranked maker's trap amid changes in the domestic and overseas business environment. Section three explains the unique collective skills of Hyundai Motor's engineers and outlines how these skills have been demonstrated in the process of developing new cars since the pilot center was built. Section four examines Hyundai Motor's horizontal manufacturing design technology engineering and how it became transformed into a vertical system

[4] See Chapters 1 and 5 for a detailed discussion of modular production.

after the establishment of the pilot center. Hyundai's engineers have established an agile production system in which the active participation of workers on the shop-floor is minimal while the engineers apply their intensive problem-solving capability to the fullest extent.

3.2 Crisis in the Business Environment and Construction of the Pilot Center

From the late 1990s, Hyundai Motor faced both internal and external growth limitations. In order to survive changes in the global business environment, Hyundai Motor should have maximized the economies of scale through continuous growth. However, Hyundai Motor's share of the US market fell due to quality problems, and its first overseas plant which was located at Bromont in Canada failed (Lansbury et al., 2006). Having failed to reach the production scale required for continued growth in the global market, Hyundai Motor became stuck as middle-ranked automakers.[5]

On the internal side, Hyundai's decision to adopt the Japanese production system entailed certain limitations. Improvements and suggestions can only be achieved by providing continued education to shop-floor workers and encouraging them to participate. However, as a latecomer, Hyundai Motor was unable to continue with systematic efforts to improve worker skills. The education and training program was not implemented consistently as it changed according to the company's circumstances and the executives in charge. In particular, it became increasingly difficult to maintain amicable labor–management relations in the 1990s after a union was established in the late 1980s. The ongoing struggle with strikes dragged on for years, intensifying the confrontation between labor and management. The distrust between labor and management only intensified when Hyundai Motor conducted major layoffs during the financial crisis in 1998.[6]

[5] At the time, the dominant discourse was that only finished carmakers with an annual production of 4–5 million units would survive in the global automobile industry. This discourse later turned out to be wrong, but it added a considerable sense of crisis to Hyundai Motor.

[6] Workers became more interested in fulfilling their rights and interests through unions than continuing to improve their skills, while remaining sensitive to employment insecurity.

By replacing the CEO in the late 1990s, Hyundai Motor responded to these changes in the internal and external business environment and engaged in a gambit to break free from the limits of a middle-ranked carmaker. Ever since the beginning of Korea's industrialization process, Hyundai Motor Group has demonstrated vertical and intensive decision-making and resource mobilization capabilities based on the concentration of authority in the head of the company, known as the *chongsu*. We conceptualized this as 'authoritarian experimentalism' in Chapter 1. After taking office in 1999, the new CEO made important decisions to change the course of Hyundai Motor. The key strategy was to promote quality innovation by putting the intensive problem-solving capacity of engineers at the forefront.

> At the time, Chairman Chung Mong-Koo advocated putting quality first and actively supported this policy. Reports of quality problems required only two things to be specified: "what is the problem?" and "who should fix it?" In other words, he designated a department head in charge of each item and had them report on the level of problem-solving each month. Chung almost always supported the opinion of the head of the quality headquarters regarding responsibility for a problem. (Interviews with a Hyundai Motor's former vice chairman K in 2020)

The construction of the pilot center lay at the heart of the new CEO's management innovation. In 2003, Hyundai Motor built a pilot center on the site of the Namyang Research and Development Center, about an hour and a half away from the headquarters in Seoul by car, located in the southern part of the Seoul metropolitan area, a separate location from the company's main plants located at the southern part of Korea. This center was in charge of integrated pilot production for the entire group. This served as a space for engineers to demonstrate their problem-solving skills in the process of developing new cars. Hyundai Motor invested 300 billion Korean won to build the center.

There are two factors that served as important catalysts for the establishment of Hyundai Motor's pilot center. The first was the rapid increase in the number of models developed and produced by the Hyundai Motor Group, increasing the need to manage pilot production in an integrated manner. Previously, pilot car production was left to domestic plants that mass-produced existing models, but as the number of models increased due to integration with Kia and the construction of overseas plants, there was a need to manage pilot production across all domestic and foreign plants at a group level. By establishing a pilot center on the site of the research institute, Hyundai Motor sought to achieve the economies of scale that could intensively solve quality and assemblability problems by integrating the pilot production functions of domestic and foreign plants.

Second, it was becoming difficult to produce pilots of new cars within mass-production plants that produced existing models. Due to the presence of strong labor unions, it had also become difficult to solve problems that arose in the development process through the skills of shop-floor workers. The time it took to produce a single new pilot car on the mass production line was equivalent to producing 7–15 mass-produced cars, which placed a burden on the plant's operations. In particular, negotiations with labor unions on the deployment of staff necessary to produce new cars had stalled, making it difficult in practice for workers to produce pilot cars. As a result, there was a growing need for the company to efficiently produce pilot cars by bypassing these conflicts with unions.

In short, Hyundai Motor built the pilot center in response to changes in the domestic and overseas business environment. This decision was made by the CEO, who sought to utilize the intensive problem-solving capability of engineers.[7] However, the construction of the pilot center did not proceed in a careful and systematic manner. The plan to build a pilot center production line in 2003. At the time, most executives

[7] The increasing role of engineers with a college degree or higher at Hyundai Motor in the late 1990s needs to be explained from the supply side of the labor market. The number of college admissions has rapidly increased in Korea along with the number of engineering college graduates since the introduction of the graduation quota system in the 1980s. As the number of engineers increased, they entered large corporations in greater numbers. These firms provided them with stable jobs and opportunities to be promoted to managers at the same time (Han & Downey, 2014). For this reason, Hyundai Motor has also recruited a large number of college graduates as white-collar workers since the early 1990s. After joining Hyundai Motor, they played a central role in both the manufacturing technology design and process technology departments based on the mechanical, electrical, and electronic engineering knowledge they gained at university. Although university graduate engineers were helped by experienced technicians with high school diplomas in the early days, they gradually came to play a central role in the process technology department and account for the majority in the staff. In other words, Hyundai Motor had enough university graduate engineers to take charge at the point they were called up to play a more active role.

opposed the construction of a large pilot center, which was unprecedented for an advanced carmaker.[8] This is an example of Hyundai Motor's leadership based on authoritarian experimentalism manifesting as risky decision-making that seeks to solve problems in a short period of time in a top-down manner rather than through a sequential decision-making process.

3.3 New Car Development and Intensive Problem-Solving Capacity

In this section, we examine how the intensive problem-solving skills of engineers have been demonstrated in Hyundai Motor's new car development process after the construction of the pilot center, especially in the pilot production stage.

3.3.1 The Collective Skills of Engineers

Hyundai Motor's catch-up to advanced automakers was made possible through the skill formation of product development engineers and manufacturing design and process technology engineers. Kim (1998) explains that Hyundai Motor and other latecomers developed their technological capabilities through a series of steps to acquire and absorb technologies. The latecomers secured manufacturing design technology first and product technology second, the reverse of the product life cycle in more advanced countries (Kim, 1998). This process can be understood by considering Hyundai Motor's strong motivation to catch-up with advanced carmakers and achieve technological independence, as opposed to passively accepting the technology provided by Mitsubishi. Expertise not included in the formal introduction of foreign technologies was gained through trial and error and reverse engineering, which refers to the act of learning design information and expertise by breaking down the machinery, product architecture, parts, and software of other companies (Hess, 2019).

[8] Mercedes-Benz's pilot line uses a cell method of consisting of 15 work stations, whereas Hyundai's pilot center uses conveyor belts in a similar fashion to the mass production line. Toyota also has a pilot center. However, because Toyota values feedback from skilled workers on the shop-floor, the pilot center is only a small facility affiliated with mass production plants (Oh, 2014).

As mentioned above, the collective skill of Hyundai Motor's engineers is mainly characterized by their intensive problem-solving ability, based on Hyundai Motor's own meta-routine. The meta-routine that Hyundai Group refers to as the 'Hyundai spirit' can be summarized as 'taking on challenges' (Kirk, 1994). This corresponds to the tendency for Hyundai Motor to choose the solution to a particular type of problem in its own way (van Driel & Dolfsma, 2010: 51). Even if there is a rather difficult task to be achieved in a short time-frame, the Hyundai spirit pursues problem-solving by concentrating resources within the business group in a pushing way. In this sense, the meta-routine of 'taking on challenges' can be thought of as intensive problem-solving capability embedded in Hyundai Motor's organizational routine. Hyundai Motor's engineers have developed an organizational culture where once a goal is set, they have considerable autonomy and do their best to achieve that goal. Hyundai Motor has set the goal of developing and producing new models that are replaced at regular intervals in each vehicle class and has accumulated internal capabilities by solving problems that arise in this process.

Hyundai Motor's engineers have built up their own skills through both continued and discontinuous interactions. This may be conceptualized as 'non-linear progress in engineer skill' as it involves the formation of skill in a non-cumulative, non-stepwise manner as opposed to a cumulative and stepwise progression. We explain this as capabilities that precede routines,[9] or the ability to improvise. Winter (2003: 993) clearly distinguishes improvisation as a capability from *ad hoc* problem-solving, arguing that the latter is neither routine nor highly patterned, while the former rests on a basis of patterned performance.

The formation of organizational capabilities in Hyundai Motor did not come about through an accumulation of routines in an orderly and sequential manner, but on a more improvisational basis (Gong et al., 2006). Attempts are made to imitate and rapidly improve foreign imported technologies through improvisational activities, and Hyundai Motor has acquired unique problem-solving capabilities in this process.

Improvisation implies the convergence of concepts, execution, and real-time planning (Miner et al., 2001), 'making a strategic plan and

[9] Routine refers to the usual sequences and methods that actors perform on a regular basis in an organizational relationship (Merriam-Webster Dictionary, 2021). In other words, routine is the systematic and collective implementation of skills.

executing it at the same time' and implies agility in the form of quickness, lightness, and nimbleness (Highsmith, 2004). Solutions may be pursued with available resources rather than optimal resources. Improvisation 'engages in spontaneous actions to create strategy within time pressure' and involves 'intuitive judgments for taking action' (Ciborra, 1999).

At Hyundai Motor, intensive problem-solving shines in the face of unexpected problems. The following quote demonstrates this:

> In the past, the A model was launched by competitor B. To respond to this, we developed the C model, and when we first developed the concept, we focused on dealing with frontal collisions. However, there is actually no such thing as a 100% frontal collision. When we tested with a prototype car, most drivers unconsciously turn the steering wheel at the last second and collide at an angle of about 40%. (...) So, just before handing the design over to the pilot, I changed the structure of one drawing which could meet the side collision criteria. The infrastructure of the model was changed after the launch date and schedule were fixed and the development of the new car was almost complete. (...) I asked if I should do it step by step, but I couldn't delay it any longer because the A model was already being released by company B. (...) To keep up with this sudden change, (...) I had to get it done in just a few days by staying up all night. (Interviews with a Hyundai Motor's executive in 2017)

In this quotation, we can see the key features of improvisation at Hyundai Motor. In the midst of the development of a new car, an important and unforeseen problem arose in product design (unexpected problem). In a short period of time where the development schedule could not be delayed (time pressure), all available resources were rapidly mobilized to solve the problem (mobilize available resources rather than optimal resources, reach convergence between planning and execution). As such, this process can be understood as conforming to the path of organizational capability-building through improvisation. Table 3.1 summarizes the characteristics of Hyundai Motor's improvisational product development model. We will elaborate on this in the next section, using the pilot center as an example.

In the 1990s, Hyundai Motor developed organizational routines by documenting its accumulated capabilities as work standards, documents that formalize organizational routines, and prescribe work processes and rules. However, work standards and work practices coexist at Hyundai

Table 3.1 Dimensions of Hyundai Motor's improvisational product development model

Minimal structuring	*Recombination*	*Responsiveness*	*Novel solutions*
• 'Making inexpensive cars with good quality' • Declarative Quality Management	• Mobilizing the experience and resources of different subsidiaries in the *chaebol* group to solve the problem • Concentrating problem-solving capabilities from throughout the company at the large-scale pilot center • Internal promotion system that mixes commitment and compulsion	• Quickly responding to unexpected problems without following a rigid predefined script • Improvised responses by the CFT as an essential coordinator • Differentiation between engineer groups in the process of product development	• Intensive problem-solving through centralized pilot center • Engineer-led integration of product design and production

Source Author's own adaptation from Kim et al. (2022)

Motor with certain differences. This manifests through the loose relationship between the former, which corresponds to 'nominal knowledge,' and the latter, which is equivalent to 'tacit knowledge.' Hyundai Motor's work practices are flexible and give broad discretion to the person in charge when it comes to detail, but are still regulated by work standards in the broader frame. Accordingly, the intuitive judgment and non-standardized improvisational capability of those in charge become important.

3.3.2 The New Car Development Process and Intensive Problem-Solving

Let's take a detailed look at how engineers solve problems in the process of Hyundai Motor's new vehicle development. As can be seen in Fig. 3.1, Hyundai Motor's functional departments work together in a parallel manner where their conflicting interests overlap during the development periods and are adjusted to successfully achieve the goals of each new car project. This is possible because the pilot center is situated in Namyang, the same location of the research and development center, so a series of activities including product design, manufacturing, and process technology preparation and pilot production can be adjusted and communicated through face-to-face contact. In this way, Hyundai Motor solves problems in the development of new cars in a simultaneous manner and has continued to shorten the time it takes from development to mass production.

The construction of a large-scale centralized pilot center served as a catalyst for new growth at Hyundai Motor. The company has implemented intensive problem-solving in the production stage of pilot cars to integrate design and production under the leadership of engineers during the pilot production phase.

As Clark and Fujimoto (1991) suggest, the integration of product design and production is a major challenge in product development in the automotive industry. However, there are a variety of ways to achieve this integration.[10] Hyundai Motor did not achieve integration between

[10] The product development organization is mostly based on a matrix structure in which functional organizations and specific project teams overlap with each other in terms of the members involved (Cusumano & Nobeoka, 1998: 64). With integration, it is necessary to consider not only integration between subsystems within the company but also the level of customer satisfaction with the product in the market. In other words, not only both

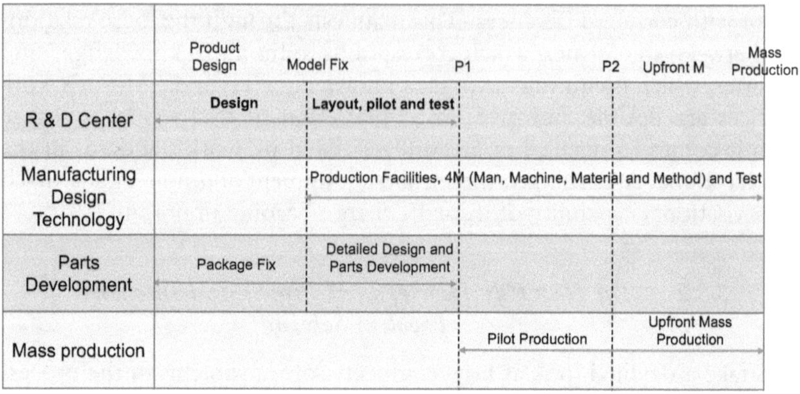

Fig. 3.1 Hyundai Motor's new car development process (*Source* Author's own adaptation from Kim et al. [2022])

product development and production through close interaction between skilled shop-floor workers and engineers, as was the case for Toyota, but through forging links between engineer-led product development and the pilot center's operations. In fact, shop-floor workers were largely excluded from the process, resulting in a successful catch-up in the low and middle automobile segments.

The intensive problem-solving capability of Hyundai Motor's engineers is fully demonstrated during the pilot production stage. Since the establishment of the pilot center, Hyundai Motor's new car development process has changed significantly in a way that focuses on solving problems during the production stage of pilot cars, prior to mass production. In the pilot car production stage, design drawings and actual products can be seen at the same time, so it is easy to identify and solve expected problems. Accordingly, the pilot center can be regarded as a key link between the R&D center and production plants. Although the pilot center is a general-purpose facility that is capable of producing cars of different models, it is equipped with welding, painting, and final assembly lines that correspond to the automobile production process, and boasts large-scale facilities similar to the production lines of mass production plants.

internal and external integration are required (Clark & Fujimoto, 1991). However, there may be various ways to achieve this integration.

The purpose of the pilot center is to identify and solve problems that could occur during mass production in advance by manufacturing pilot cars during the development process. At the pilot center, pilot cars for new car models are produced on two production lines.

At the beginning of the center's establishment, engineers with experience in the design of cars and manufacture of prototype cars, production design and process technology, and parts purchasing areas were recruited from Hyundai Motor Group's plants to work at the pilot center, but the proportion of newly hired engineers steadily rose from that point onward. A total of 605–625 employees work at the pilot center, including 70–80 middle managers in charge of secretaries in the Cross-Functional Team (CFT), and shop-floor workers who are directly in charge of pilot car production.

Figure 3.2 shows how the pilot center's CFT has intensively solved problems in the pilot car production stage. First, the chief team manager of the CFT is the head of the pilot center, and middle managers from the pilot center, advanced manufacturing design center, and proto car shop work as secretaries who oversees the task of intensive problem-solving. The pilot center and proto car shop's secretary is in charge of quality, while the secretary of the advanced manufacturing design center is responsible for assemblability issues. Second, the CFT is a sort of matrix organization, composed of functional units for subsystems such as chassis, exteriors, interiors, and moving, as well as departments in charge of integration and coordination in specific project units including the pilot center, proto car shop, advanced manufacturing design center, production, quality, and the research institute. Each engineer from the pilot center, proto car shop, and advanced manufacturing design center serves as the secretary and coordinates shared work with each department. Representatives from the departments in charge of the secretaries reside at the pilot center, but those from other departments only come to the pilot center and participate in meetings when it is necessary to resolve issues. Each full-time secretary of the CFT is in charge of projects for 2–3 vehicle types at the same time in subsystem units. Third, the CFT's objective is to identify expected quality and assemblability problems in advance at the pilot car production stage and resolve them before the car enters mass production through meetings between the representatives in charge.

The pilot center produces an average of 100 pilot cars per new product at the P1 and P2 stages. Both P1 and P2 have mass-production molds to produce cars under conditions that resemble mass production, and each

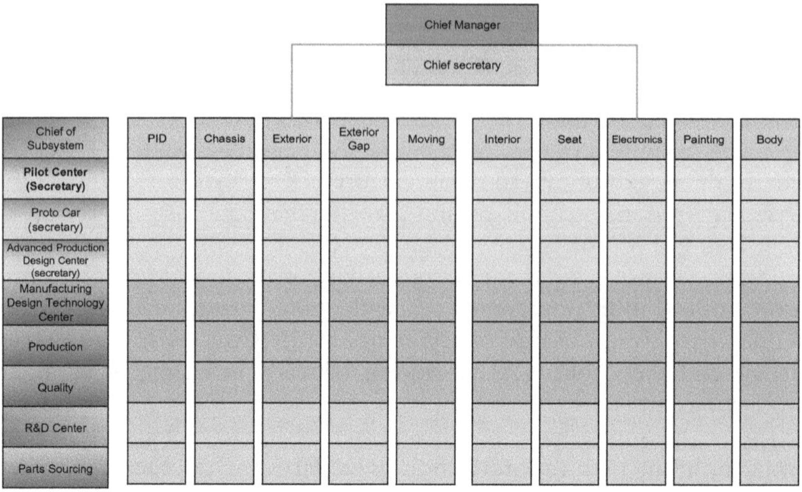

Fig. 3.2 Organizational chart of the cross-functional team at the Hyundai Motor's pilot center (*Note* PID = Pre-inspection Delivery. *Source* Author's own adaptation from Kim et al. [2022])

stage takes an average of one month. The CFT activity of a vehicle is active during P1 and P2 and finished in two weeks after mass production. The system identifies an average of 1000–1500 problems per vehicle type in the pilot production stage and seeks to determine the cause and implement improvements. This includes quality and assembly problems that could occur during mass production, such as surface treatment and heat treatment. The goal of CFT is to mediate conflicts between departments and solve problems, including when external parts suppliers are connected to the problems identified. Once an issue is identified, the CFT asks the corresponding department to solve the problem by changing the design, improving equipment or parts, or promoting the standardization of assembly with the same structure of different car models.

As an example, imagine that the window wiper linkage of a new product makes a noise when the wiper moves downward. The CFT will identify the cause and then modify the design data, amend errors in the manufacture of parts, or improve the error range during parts assembly.

Let's explain the role of the CFT secretary in more detail. The interests of Hyundai Motor's functional departments that participate in new car development are often not in alignment. If a noise problem occurs during the production of a pilot car, it is necessary to hold a CFT meeting to determine whether the main cause comes from the research institute in charge of design, the parts suppliers, or the plant in charge of assembling the finished product. If it is difficult to reach an agreement, the CFT secretary plays an important role as a coordinator and mediator. When the secretary is asked to resolve a disagreement between the heads of each functional department, their decision is accepted in most cases. Since the secretary has built up experience in solving dozens of new car problems per year as a member of the pilot center, they oversee the problem-solving process. CFT secretaries aim to meet the quality assurance standards that correspond to work standards within a fixed period of time during the pilot production stage, but when a satisfactory resolution cannot be reached in practice, they can improvise or exercise discretionary power to solve the problem.

For example, if a quality issue with a part occurs when a part supplier has failed to develop a mass production mold, there is a case where the pilot production stage is skipped to the next stage in consideration of the mass production schedule even though the quality assurance standard has not been met, although this is rare. In this case, the secretary would not be held responsible for personnel affairs even if a problem manifests at a later date. The CFT secretary's exercise of discretion is based on personal problem-solving experience within the CFT organization, and the organizational culture tolerates the use of such discretion. In this sense, each engineer's improvisational capability is exercised on a collective rather than individual level.

Take one more example of improvisation in the stage of pilot production. When mounting a seat belt in the interior of a pilot car, it is difficult to accurately check a torque value because it is not possible to accurately check how the seat belt is fastened to a carpet or car body. Although the torque value of the seat belt mounting is fixed by engineering, since there is a lot of dispersion and the condition changes according to the season, there is a lot of difference from the design drawing. It is in these circumstances that the CFT secretary engineers' ability to improvise comes into play. Since the exact torque value for a specific situation is not determined by regulations, engineers are responding in an improvisational way.

Let's look at an example of solving an assemblability problem. When mounting a roof rack on a vehicle, it is necessary to alter the design since the structure of the assembly line workspace is different in each plant. For example, the team has to reduce the number of parts and mount them all at once when the work space is smaller, or reduce the length of the bolts or enlarge the holes to facilitate mounting the rack when there is a lower ceiling. The CFT secretary also plays an important role in resolving these kinds of assemblability issues, and is able to exercise discretionary power as a coordinator and mediator if disagreements cannot be resolved.[11]

During the pilot production stage, engineers' problem-solving capabilities are also demonstrated in their relationships with external partner parts suppliers. Engineers from these parts suppliers have participated as collaborative guest engineers with careful deliberation and deep discussion in almost all stages of product development at Hyundai Motor, from design to preparation for mass production. In the pilot production stage of a new product, these parts makers participate by supplying molds for mass production. CFT engineers discover and then solve problems in design, quality, and assemblability, with manufacturing pilot cars with the molds ready to mass-produce supplied by these parts suppliers. As the technological capabilities of these parts makers improve, the problems caused by the poor quality of parts and components tend to decrease. As the modularization of parts and components progresses, the responsibility of these parts suppliers for quality improvement and cost has more increasingly widened than before.

In the course of the pilot center's work, how are differences of opinion between functional departments handled if they cannot be resolved autonomously on a horizontal level? Most issues arising from the pilot production phase are resolved at CFT meetings, but 5–10% of quality issues remain unsolved due to disagreement or non-cooperation from the functional organization.[12] For example, 1450 quality problems occurred in the development of one product, of which 70 unresolved issues were brought to the chief CFT meeting, a weekly meeting attended by heads of department that is designed to deal with issues that cannot be resolved in individual CFTs. Because upper-level executives have a lot of authority,

[11] In the pilot center's problem-solving process, quality and assemblability issues are considered to have a 50:50 ratio, but this can be seen as a 70:30 ratio when considering their actual importance.

[12] Problems need to be discovered and resolved, but when it would be politically awkward to point out a large number of problems in a specific department, there are cases in which the nature of problems are concealed or responsibility is not made explicit (Interviews with a Hyundai Motor executive in 2014).

they are able to adjust the development schedule for a new car or make requests to other departments regarding issues with design or parts suppliers.

In short, Hyundai Motor has solved quality and assemblability problems in the new car development process through an intensive problem-solving framework in the pilot production stage. When it is difficult to autonomously solve problems among the heads of each functional department, the full-time secretary at the pilot center makes decisions on an *ad hoc* and improvisational basis by drawing on their problem-solving experience. In most cases, the secretary's decision is accepted, but if a resolution cannot be reached, it is brought to the chief CFT meeting for upper-level personnel to solve.

Hyundai Motor also fixes problems in the production process by having shop-floor workers participate in the pilot stage. However, the role of shop-floor workers has declined over time. Pilot vehicles are produced by around 300 shop-floor workers at the pilot center. The pilot center initially launched with 30–40 skilled workers from domestic plants, but newly recruited shop-floor workers have accounted for the majority since then. In the early days, these employees were actively involved in identifying and pointing out problems with pilot cars, but they are now primarily engaged in simple tasks such as assembly and have limited interaction with the engineers. The low-skilled nature of workers has been strengthened, and engineers are currently responsible for most of the problem-solving and design changes.

3.3.3 Comparison with Toyota

How is Hyundai Motor's new car development process different from that of Japanese automakers such as Toyota? First of all, while Hyundai Motor values problem-solving in the pilot production stage, Toyota solves most of the problems in the product design stage before entering pilot production. At Toyota, most of the problems that may occur in new vehicles are resolved through frequent changes in the product design stage. In contrast, Hyundai Motor's large pilot center makes up for the lack of problem-solving capabilities in the product design stage through intensive problem-solving. In this sense, the core of Hyundai's organizational competence and competitiveness lies in its problem-solving capacity through the pilot center.

Second, the two companies differ greatly in the development of their organizational capabilities. According to Whitney et al. (2007), Toyota favors standardizing production processes and design, where all essential procedures are recorded as standards in the production process. Engineers choose their product data and design methods and perform routine work based on established standards. These standards are rigorously updated by the 'Evaluation Group,' which consists of several engineers. This process can be described as incremental and cumulative innovation based on clear standards.

On the other hand, Hyundai Motor's growth in organizational capacity does not proceed in a cumulative and staged fashion. There are often important problems that appear in the pilot center that were not predicted in product design. Available resources are mobilized to solve these problems, and the intuitive judgment and improvisational ability of engineers play a significant role in this process. This more closely resembles capacity building through organizational improvisation than fixed standards and routines.

Finally, Hyundai Motor and Toyota also differ in their decision-making structures. In new car development, it is crucial to improve the expertise of each functional department and manage their competing interests in an integrated manner. To this end, both Toyota and Hyundai Motor have chosen a matrix structure for new car development.

However, the decision-making structure of the matrix is different in each company. At Toyota, a matrix organization with agents from each functional department leads the development of new cars under a project manager (PM) or chief engineer (Chief Executive). The PM has the authority to set-up product concepts, adopt key technologies, write the main specifications, manage sales targets, manage costs and profits, and select key members. They strive to achieve integration in new product development while also frequently carrying out direct talks with engineers from each functional department. At Toyota, the PM has the same level of standing as the head of each functional department, and conflicts or disagreements are resolved through direct talks with the functional department head (Jo & Oh, 2020: 158–159).

As discussed above, Hyundai Motor also has a matrix organization called CFT in the development of new cars. However, the CFT is an entity with multiple engineers and multiple secretaries, but no single

PM holds the same level of authority as at Toyota.[13] When differences between functional departments appear, Hyundai Motor first attempts to solve the problem through a general CFT meeting, and if the problem is not resolved there, the CFT then finally reports the problem to the chief executive officer to resolve the issue.

Finally, Hyundai Motor and Toyota also differ in the relationship between manufacturing design technology and process technology. At Toyota shop-floor workers actively participate in improvement activities. The division of work between process and manufacturing design technology is horizontal, with shop-floor workers encouraged to make suggestions (Adler et al., 1999; Shibata, 2009: 1902–1905). On the other hand, Hyundai Motor's intensive problem-solving leans more toward manufacturing design technology in the advanced or front-end process. Shop-floor workers play a lesser role in improvement activities, and thus process technology engineers are more burdened with most of the responsibility for making improvements. Table 3.2 summarizes the characteristics of the new car development process at each automaker.

3.4 The Vertical Restructuring of Production Engineering

The construction of the pilot center enabled Hyundai Motor to establish a production engineering system that includes both manufacturing design technology and process technology and is distinct from the Japanese production system. This section compares Hyundai Motor's horizontal production engineering in the early days of industrialization with the vertical production engineering that emerged after the pilot center.

3.4.1 Horizontal Production Engineering at Hyundai Motor

In the early days of industrialization, Hyundai Motor attempted to absorb the Japanese production system through a partnership with Mitsubishi. As a latecomer in the automobile industry, Hyundai Motor believed that imitating the production systems of advanced automobile companies

[13] This is partly because Hyundai Motor does not have an executive with extensive experience to cover the entire new car development process, but also reflects the characteristics of Hyundai Motor's organizational culture, which gives a greater voice to functional organizations (Jo & Oh, 2020: 159).

Table 3.2 A comparison of the new car development process: Hyundai Motor vs. Toyota

	Hyundai Motor	*Toyota*
Product development model	Improvisational model	Integrated model
Problem-solving	Making up for unaddressed issues in the product design phase with intensive problem-solving capabilities in the pilot production phase	Solving most problems during the product design phase
Organizational capacity-building	Innovation through organizational improvisation, and intensively solving problems by mobilizing all available resources to solve critical problems that were not anticipated in product design	Gradual and cumulative innovation based on standards, i.e. performing routine tasks based on established standards, which are updated very strictly by the 'Evaluation Department' (Whitney et al., 2007)
Organization of new car development	A matrix organization called CFT solves problems through general CFT meetings when disagreements occur between functional departments. If the problem is not solved here, a multi-layered vertical information processing system will resolve it by reporting it to the CEO	The PM has strong authority as a leader in new car development and leads the integration of new product development. If necessary, conflict and confrontation are resolved through direct talks with the functional department head
Relationship between manufacturing design technology and process technology engineers, and shop-floor workers	A higher proportion of manufacturing design technology engineers is in charge of front-end or advanced processes, while the role of process technology engineers is limited to a focus on back-end processes. The latter is responsible for most improvement activities, while shop-floor workers play a lesser role	Horizontal division between manufacturing design technology and process technology engineers, with active participation by shop-floor workers in improvement activities

Source Author's own adaptation from Cunha and Gomes (2003), Whitney et al. (2007), and Jo and Oh (2020)

with a confirmed competitive advantage would be the best strategy to reduce trial and error. Hyundai Motor adopted the Japanese production system through various unofficial channels such as business trips, industrial training, and the use of retirees from Japanese companies as advisors, as well as official channels of technical cooperation with Mitsubishi (Hyundai Motor Company, 1987).

Hyundai Motor's core research center was established in the early days of Korea's industrialization and was located within the Ulsan plant. At that time, the functions of product development and production technology engineering appear to have been integrated within the research center. Engineers and other employees in the research center and plant met frequently to communicate and solve problems in the process of new car development (Amsden, 1989). This horizontal production engineering mimicked the production system of Japanese automakers.

What is noteworthy in production engineering during the early stages of industrialization is the role of high school graduate technicians as process technology engineers. As Hyundai Motor pushed ahead with the development and production of its own model, the company had to establish a mass production system in a short period of time and improve its problem-solving capability to meet the level of quality required for the export market. During this period, technicians from technical high schools played a central role in production engineering. With university graduate engineers in short supply at the time, high school graduate technicians with on-the-job experience were recruited as process technology engineers to fill the gap.

As Hyundai Motor gradually shifted to a mass production system, production engineering became more systematic. Since it was important to have experience on the shop-floor for the development of new car models and mass production, a number of excellent high school graduate shop-floor workers were recruited as process technology engineers. Excellent technicians trained at specialized technical high schools began to flock to Hyundai Motor from the 1970s, and they have played an important role in reverse engineering, absorbing, and assimilating technological knowledge from Japan in the early days of industrialization and applying this to the shop-floor.

Hyundai Motor selected excellent technicians and conducted a year of long-term education and training at the time which was unconventional at that time. The company actively promoted improvements and suggestions through Quality Control (QC) meetings. With the establishment of the process technology department in 1983, these technicians started to bring about gradual innovation in production technology. Their main task was to solve problems in the development of new cars and production processes by communicating and cooperating with the research center and manufacturing technology engineers. Because they were from the shop-floor, they had a wealth of experience and were able to cooperate smoothly with shop-floor supervisors and workers.

Before the establishment of unions, there was a clear hierarchy between university graduates and high school graduates in the early stages of Korea's industrialization, and discrimination was also rife. However, these groups were able to cooperate with each other relatively effectively at Hyundai Motor due to the common goal of catching up with advanced competitors. Technicians working as process technology engineers communicated problems identified on the shop-floor with manufacturing design technology engineers, modified molds and/or equipment, and where necessary communicated with those in charge of the research center to make design changes, thus improving Hyundai Motor's organizational capabilities.

In short, design, manufacturing design technology, process technology, and shop-floor workers all cooperated in a horizontal structure of new car development and mass production during the early stages, establishing an ethos of shop-floor-oriented production engineering with a focus on process technology. This is similar to the integrated organizational capabilities of Japanese automakers, including Mitsubishi and Toyota, which Hyundai Motor sought to imitate and catch up with.[14]

3.4.2 *Integrated Production Engineering in Japan*

In Japanese automakers, production engineering is integrated through horizontal cooperation between manufacturing design technology engineers and process technology engineers. The former are involved in work

[14] However, since this period was prior to the establishment of labor unions, it is necessary to note that the shop-floor was under the authoritative and discretionary control of supervisors on the shop-floor. They exercised their authority in performance reviews linked to promotions and various bonuses in a discretionary manner (see Chapter 4). Under this type of management, workers had no choice but to passively engage in simple, repetitive labor (Jo, 1993).

related to product design from the beginning and prepare the production line for mass production, while the latter delivers feedback from the shop-floor to manufacturing technology design engineers through interaction with shop-floor workers. Horizontal personnel exchanges are also conducted frequently between both groups. As the level of automation and informatization increases, the nature of production engineering has changed, and the proportion of engineers with a college degree or higher has also risen in Japan. It is said that around 70% of productivity gains come from improvements made by production engineers, and only 30% are the result of improvements made by workers and supervisors on the shop-floor (Shibata, 2009: 1901).

However, the tradition of valuing the shop-floor continues to this day. Production technology engineers experience work on the shop-floor for more than six months during the new hire period. At Japanese automakers, the group that corresponds to high school graduates at Hyundai Motor is technicians from a technical high school background. These are skilled workers who have completed three years of technical high school and two years of technical college and are trained at 60 or so educational institutes across Japan. These technicians play a central role in production engineering by engaging in highly skilled work such as facility maintenance and quality control. Although initially hired as shop-floor workers, their skills are continually improved through intra-firm education and training such as the skill qualification system and professional skill system. The personnel system also offers a path for them to be promoted to white-collar managers or shop-floor supervisors such as foremen (Jo & Oh, 2020). Workers in general also communicate smoothly with both manufacturing design technology and process technology engineers, and their involvement on the shop-floor contributes to improvements and higher quality.

3.4.3 *The Establishment of Hyundai Motor's Agile Production System*

Following the reorganization of the late 1990s, Hyundai Motor's production engineering is now distinct from the Japanese production system, moving from a horizontal structure to a more vertical hierarchy that connects manufacturing design technology engineers, process technology engineers, and workers.

While manufacturing design technology engineers arrange the equipment, tools, and parts for mass production after a new vehicle is developed, process technology engineers are in charge of technical methods and work organization, as well as process improvement. The vertical division of labor between them has become more pronounced in the process of new car development. As the number of vehicles increases, manufacturing design technology engineers have a steady stream of work as they continue to develop new vehicles throughout the plant. On the other hand, as process technology engineers are in charge of new car development only once every 3–4 years at the plant they work at, their role is more intermittent until the next development cycle. In other words, since process technology engineers have comparatively less experience in vehicle development, their main function is to provide feedback to the manufacturing design technology engineers on existing mass-produced vehicles. Their degree of involvement in design modifications has also declined. In addition, personnel exchanges between engineers in the two departments are now limited to exceptional cases when the number of applicants match both sides and are conducted on a small scale.

Nevertheless, process technology engineers do play a central role on the shop-floor. They, with some help of supervisors such as team and group leaders on the shop-floor, organize the workflow and take charge of quality issues. These process technology engineers often suffer from excessive workloads because they have to take care of labor management in addition to their own work. When work is organized through production engineering, most workers perform standardized tasks in a simple and repetitive manner. Although they are in charge of their respective processes, they tend to be less interested in skill formation or improvement activities. Job rotation between similar processes within the same team is sometimes carried out to reduce monotony, but there is no systematic multi-skilled work beyond each team unit (see Chapter 4, for more detail).

With the construction of the pilot center, Hyundai Motor established an agile production system that differs from that of Japan. In this system, the skills and involvement of shop-floor workers play a minimal role, while the intensive problem-solving capabilities of engineers, including external

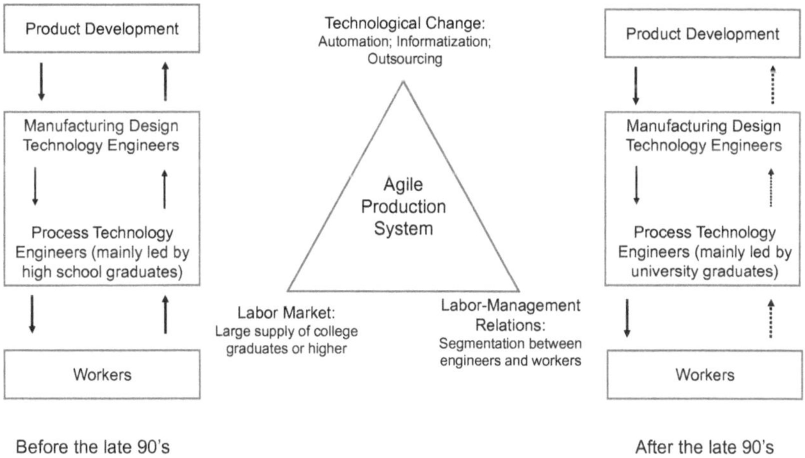

Fig. 3.3 Changes in Hyundai Motor's organizational capabilities in production engineering (*Note* Solid lines refer to active action, while dotted lines denote passive action. *Source* Author's own creation)

suppliers, are more central. Engineers are in charge of the planning function to the fullest extent of their professional abilities, and workers on the shop-floor are only responsible for simple execution. In this way, the system is characterized by a limited level of collaboration between engineers and parts suppliers, while largely excluding shop-floor labor. Figure 3.3 summarizes how the organizational capabilities of Hyundai Motor's production engineering have changed over time.

3.5 Conclusion

Since the late 1990s, Hyundai Motor's production engineering has been restructured from horizontal production engineering akin to Japanese carmakers such as Toyota to a system of vertical production engineering with engineers at the forefront. This has made a positive contribution to Hyundai Motor's success in breaking free from the 'middle-ranked carmaker's trap' that firms often face in developing countries. In order to overcome the sluggish overseas market and confrontational labor relations, the newly inaugurated CEO built a pilot center through the centralized and vertical decision-making of authoritative experimentalism as described in Chapter 2. This was an innovative move to leverage the collective skills of engineers as a meta-routine of the company. From this point onward, Hyundai Motor has carved a development path clearly that definitely differs from the Japanese production system. The collective skill of Hyundai Motor's engineers is an improvisational and intensive problem-solving ability in which improvisation takes precedence over routine. In other words, Hyundai Motor's engineers have the ability to achieve project goals by solving problems in an improvised way once a certain goal is set.

Since the construction of the pilot center, engineers are able to demonstrate their intensive problem-solving capabilities in the pilot production stage of new car development by identifying and resolving problems in the design drawings and manufacturing process at the same time. The CFT staff working as secretaries in the pilot center resolve issues arising during the production process on an improvisational basis by exercising their discretion while discussing issues with the persons in charge of each functional department and external parts suppliers. Problems that cannot be resolved through the process are reported to high-level executives.

The vertical production engineering found in Hyundai Motor's new car development process is different from that of Toyota, where a project manager acting as the chief engineer oversees all stages and solves problems horizontally with each functional department. Toyota is able to detect and solve many problems in advance during the design stage, whereas Hyundai Motor intensively finds and solves many problems in the pilot production stage in comparative terms. The ability of engineers to solve problems is highlighted more clearly because Hyundai Motor does not expect active participation from workers due to confrontational labor–management relations. In short, the agile production system that Hyundai Motor established with the construction of the pilot center is characterized by intensive problem-solving capability among engineers and the skill-saving work organization of production workers, which will be discussed further in Chapter 4.

References

Adler, P. S., Goldoftas, B., & Levine, D. I. (1999). Flexibility versus Efficiency? A Case Study of Model Changeovers in the Toyota Production System. *Organization Science, 10*(1), 43–68.

Amsden, A. (1989). *Asia's Next Giant: South Korea and Late Industrialization*. Oxford University Press.

Amsden, A., & Hikino, T. (1994). Project Execution Capability, Organizational Know-how and Conglomerate Corporate Growth in Late Industrialization. *Industrial and Corporate Change, 3*(1), 111–147.

Ciborra, C. U. (1999). Notes on Improvisation and Time in Organizations. *Accounting, Management and Information Technology, 9*(1), 77–94.

Clark, K. B., & Fujimoto, T. (1991). *Product Development Performance: Strategy, Organization, and Management in the World Auto Industry*. Harvard Business School Press.

Cunha, M. P., & Gomes, J. F. S. (2003). Order and Disorder in Product Innovation Models. *Creativity and Innovation Management, 12*(3), 174–187.

Cusumano, M. A., & Nobeoka, K. (1998). *Thinking Beyond Lean: How Multi Project Management is Transforming Product Development at Toyota and Other Companies*. Free Press.

van Driel, H., & Dolfsma, W. (2010). Imprinting, Path Dependence and Metaroutines: The Genesis and Development of the Toyota Production System. In G. Schreyögg & J. Sydow (Eds.), *The Hidden Dynamics of Path Dependence: Institutions and Organizations* (pp. 35–49). Palgrave MacMillan.

Fujimoto, T. (2001). *Introduction to Production Management*. Nohon Keizai Press. (In Japanese).

Gong, Y, T., Baker, T., & Miner, A. S. (2006). *Capabilities and Routines in New Organizations: Evidence from the Field* (Working Paper). School of Business, University of Wisconsin.

Han, K., & Downey, G. L. (2014). *Engineers for Korea*. Morgan & Claypool Publishers.

Hess, B. (2019). What Is Reverse Engineering and How Does It Work? Link available at https://astromachineworks.com/what-is-reverse-engineering/

Highsmith, J. (2004). *Agile Project Management: Creating Innovative Projects*. Pearson Education.

Hyundai Motor Company. (1987). *20 Years of Hyundai Motor Company: 1967–1987*. Hyundai Motor Company. (In Korean)

Jo, H. J. (1993). *Strategic Option of Korean Automobile Industry*. Baeksan. (In Korean)

Jo, H. J., & Oh, J. (2020). *The Development Process of New Cars for Overseas Markets: A Comparison of Toyota and Hyundai*. University of Ulsan. (In Korean).

Kim, C., Jeong, J. H., & Jo, H. J. (2022). Improvisational Intensive Problem-Solving Capability: The Case of Hyundai Motor's New Car Projects. *Journal of Asian Sociology, 51*(4), 323–354.

Kim, L. (1998). Crisis Construction and Organizational Learning: Capability Building in Catching-up at Hyundai Motor. *Organization Science, 9*(4), 506–521.

Kirk, D. (1994). *Korean Dynasty: Hyundai and Chung Ju Yung*. DRT International.

Lansbury, R. D., Kwon, S.-H., & Suh, C.-S. (2006). Globalization and Employment Relations in the Korean Auto Industry: The Case of the Hyundai Motor Company in Korea, Canada and India. *Asia Pacific Business Review, 12*(2), 131–147.

McQuaid, R., & Lindsay, C. (2005). The Concept of Employability. *Urban Studies, 42*(2), 197–219.

Merriam-Webster Dictionary. (2021). Link available at https://www.merriam-webster.com/dictionary/dictionary

Miner, A. S., Bassoff, P., & Moorman, C. (2001). Organizational Improvisation and Learning: A Field Study. *Administrative Science Quarterly, 46*(2), 304–337.

Nobeoka, K. (2007). The Stack of Organizational Capability: Technology that Cannot Be Imitated. *Organizational Science, 40*(4), 4–14. (In Japanese).

Oh, J. (2014). *Product Development Process at Toyota: A Research Note*. Meiji University. (In Korean).

Shibata, H. (2009). A Comparison of the Roles and Responsibilities of Manufacturing Engineers in Japan and the United States. *The International Journal of Human Resource Management, 20*(9), 1896–1913.

Whitney, D. E., Heller, D. A., & Fukuzawa, M. (2007). *Production Engineering as System Integrator? A Research Note based on a Study of Door Engineering and Assembly at Toyota Motor Corporation* (University of Tokyo MMRC Discussion Paper No. 169).

Winter, S. G. (2003). Understanding Dynamic Capabilities. *Strategic Management Journal, 24*(10), 911–995.

CHAPTER 4

Work Organization and Shop-Floor Workers—Flexible Automation, Skill-Saving, and Segmented Labor

4.1 Introduction

This chapter examines the characteristics of intra-firm work organization and labor in Hyundai Motor's agile production system. In the previous chapter, we discussed the skill formation of engineers in the process of developing Hyundai Motor's new vehicles. As the pilot center solves most problems through intensive and improvisational problem-solving by engineers, the mass production process in Hyundai Motor's production system now consists of a work organization aimed at faithfully carrying out standardized production instead of pursuing innovation. In contrast to engineers, the shop-floors need little skill. Accordingly, we conceptualize Hyundai Motor's work organization at the mass production site as 'skill-saving work organization.'

Flexible production has long been a global trend in the automobile industry. In line with this, Hyundai Motor developed a flexible production system as early as the 1990s. But how does this flexible production combine with skill-saving work organization?

In literature theorizing about flexible production systems such as flexible specialization, lean production systems, and new production concepts, flexible production has been discussed in connection with a reduced vertical division of labor, autonomy, and skill improvement in work organization (Kuhlmann & Schumann, 1997; Piore & Sabel,

H. J. Jo et al., *Agile Against Lean*,
https://doi.org/10.1007/978-981-99-2042-6_4

1984; Womack et al., 1990). From this point of view, Hyundai Motor's skill-saving work organization appears contrary to flexible production.

However, re-specialization and skill-building is not the only way to achieve flexibility. Conversely, flexibility can also be achieved by standardizing and simplifying the work process and marginalizing labor. For example, technologies and expertise are applied to the concept stage such as planning and design and embodied in the machine, while the mass production process can be standardized and simplified so that workers can easily perform their jobs and adapt quickly and flexibly to the situation at hand. This can be conceptualized as 'flexible standardization,' where skills and specialized abilities are less important and labor is largely viewed as a necessary cost that should be reduced as much as possible (Kim, 2011). At Hyundai Motor, flexible production has been carried out in this way. The company achieved flexibility on the shop-floor by applying flexible automation technology, while the work process has been simplified as much as possible to reduce deviations caused by human factors. This skill-saving work organization combined with flexible automation technology has served as the basis for replacing full-time labor with machines or more insecure work arrangements.

In the second section of this chapter, we discuss the characteristics of Hyundai Motor's production technology and work structure before the 2000s, which formed the foundation for skill-saving work organization. Section three examines the formation process of Hyundai Motor's production system. This covers Hyundai Motor's modular production, flexible automation and informatization, working conditions under the skill-saving work organization, and a multi-layered quality management system that compensates for the lack of skill on the shop-floor. Section four looks at the contribution of labor–management relations in creating the skill-saving work organization system and describes the characteristics of the resulting labor market structure. Section five demonstrates the specificity of Hyundai Motor's production system through comparison with Toyota's. We argue that Hyundai Motor's current production system has evolved due to a combination of structural and behavioral variables that the company faced in the process of system restructuring in the automobile industry, and is not just a variation of Fordism or Toyotism. Finally, section six summarizes the discussion.

4.2 Background to Skill-Saving Work Organization—Before the 1997 Financial Crisis

4.2.1 Development of Production Technology and Labor-Replacing Automation

The origins of Hyundai Motor's production system go back to the Pony project, the development of the company's first independent model. After failing to forge a partnership with Ford in the early 1970s, Hyundai Motor began the development of its first proprietary model, the Pony, thanks to the Korean government's 'Long-Term Automobile Industry Promotion Plan.' The Hyundai Group supported the Pony project by dispatching engineers from its affiliates to Hyundai Motor and generously providing management resources. After the success of the Pony project, the company began to develop a production system suited to its own models, thereby establishing itself as the leading enterprise in the Korean automobile industry (Hyundai Motor Company, 1987). In the 1980s, Hyundai Motor actively adopted automation when it built a plant that produces Excel, an export strategic vehicle model. The motivation for aggressively pursuing automation was the need to maintain uniform quality. In addition, the need to reduce personnel costs by replacing labor also served as an important motivation for automation. As an illustration, if the price of a robot at the time was around KRW 50 million won and had a depreciation period of three years, it would be economical for one robot to replace two workers on annual salaries of KRW 10 million won.

At the same time, Hyundai Motor has also pursued informatization to support the development of production technology. The company has been partially computerized in materials, production, and accounting since the late 1970s, but as the number of units produced grew to 100,000 a year and assembly specifications became more complicated, there was a need to systematically implement production management through informatization. Hyundai Motor's engineers, who had delivered production plan information to the back-end process by bicycles, began to introduce information systems for systematic production management in the mid-1980s, when the number of productions increased to 300,000 a year.

From this point on, Hyundai Motor introduced the 'Bill of Materials' (BOM) that classifies and manages parts into 250–400 groups, as well as the 'Material Requirement Plan' (MRP) system based on BOM, which calculates parts requirements according to production plans (Jo & Cho, 2012). During this period, Hyundai Motor also developed the 'Assembly Line Control' (ALC) system that directs work in the final assembly process based on the BOM and MRP. In short, the company achieved more systematic production management by confirming the specifications of complex and bulky parts such as engines, suspensions, seats, and bumpers through an information system that inputs them into the final assembly line (Hyundai Motor Company, 1987: 460).

In the 1990s, Hyundai Motor built a new plant in Asan. This was some distance from Ulsan, where the existing plant was located. In addition to uniform quality, the new plant promoted automation in order to replace avoidable processes such as handling heavy materials and reduce paint costs. The Asan plant uses a 'green-field strategy' to avoid militant labor unions, and automation is part of that strategy. Hyundai sought to suppress the transfer of workers from the Ulsan plant as much as possible to prevent the working atmosphere and labor–management relations from spilling over. The majority of workers were young, local, and inexperienced, and appointed through a recommendation system. At the same time, Hyundai Motor tried to reduce the influence of labor unions by minimizing the number of employees through automation (Chung, 1997).

It was after the 1990s that Hyundai Motor ramped up the development of flexible production technology. From this point on, the company promoted flexible automation as a way to efficiently respond to changes in market demand. Flexible automation involves producing multiple models on the same production line through general-purpose equipment and software. Hyundai Motor has gradually advanced this technology in both local and overseas plants in the areas of production processes, transportation between processes, quality, and production information systems (Interviews with a Hyundai Motor executive in 2013).

Table 4.1 shows how automation has progressed at Hyundai Motor, demonstrating that automation has advanced to a higher level with the establishment of each new plant.

4.2.2 *A Failed Attempt to Build Workplace Skills*

Despite the dominant trend of replacing labor with automation, Hyundai Motor did attempt to systematically build and utilize the skills of shop-floor workers. In the 1990s, the company introduced a skill formation

Table 4.1 Hyundai Motor's automation by plant

	Ulsan No. 1 Plant (late 1970s)	*Ulsan No. 3 Plant (late 1980s)*	*Asan Plant (mid-1990s)*
Stamping shop	Tandem Press: Manual work	Transfer Press: Automation by linking both loading and unloading	Double Stamping by Transfer Press: Loading on pallets automated and automatic guided vehicles (AGV) used for moving
Welding shop	Manual spot welding and manual transport (use of auxiliary equipment)	Full-scale introduction of welding robots and shuttle transfer (repeating lifting, carrying, down & return)	Introduction of robots for welding, measurements and inspections and adoption of flexible body lines (FBL) in all processes
Painting shop	Manual spraying: Transported by a conveyor belt and dipped into the paint tank	Cation electrodeposition coating (anti-rust) and spray painting by rotating at high speed (mini bell method with more than 80% transfer efficiency)	Robots used for painting, spraying and sealer work, and interior painting (low investment efficiency)
Final Assembly shop	Conveyor belt	Conveyor belt and shuttle transfer introduced to line 2 in No. 3 plant	Conveyor belt and equipped trams following a certain track (Stop during work; stop where necessary, and work height to be adjusted) 10% based on man-hours

Source Author's own adaptation from Hyundai Motor's internal documents, 2012

system on two occasions with the aim of adopting a flexible work organization in line with flexible production technology. The first attempt was the 'skill qualification system' in the early 1990s, which encouraged the formation of skills through reorganization. The main features of this system were gradually improving the level of skill through on-the-job training (OJT) and off-the-job training (Off-JT) to help production workers earn promotions to team leaders or technicians. However, the new system was thwarted by opposition from unions, which argued the system could exacerbate individual competition among union members and weaken unity (Jo & Lee, 2008: 245).

The second attempt came about in 1997 during the Asian financial crisis. Hyundai Motor sought to improve the skills of production workers through education and training on spare workers who increased rapidly due to a drop in the utilization rate. The core component was the 'Training Road Map (TRM)' program as an extension of the previous 'skill qualification system,' aimed at systematically transferring knowledge embedded in the form of 'tacit knowledge' from senior to junior staff. Such an attempt seemed likely to be met with conditions to inject surplus staff into education and training in the wake of the financial crisis, but it was thwarted by fierce labor–management conflicts and the replacement of top executives in the same year (Jo & Lee, 2008: 245–246).

After the failure of these two attempts, Hyundai Motor chose to develop a technology-oriented and skill-saving production system that minimizes dependence on shop-floor worker skills.

4.3 Full-Scale Flexible Automation and Skill-Saving Work Organization in the 2000s

After establishing an exclusive domestic system in Korea with the acquisition of Kia in 1999, the Hyundai Motor Group, sought to ramp up Hyundai Motor's production system by pursuing an aggressive management strategy. In this process, the company actively used flexible automation technology in the workplace to achieve both cost reductions and flexibility.

4.3.1 Development of Modular Production

Modular production played an important role in the establishment of these systems. Hyundai Motor introduced modular production later than other automakers in Europe and the United States, but has applied it at a higher level than any other automaker in the world. Modularization is defined as the process of building complex products or processes into smaller subsystems that function together as a whole but can be designed independently (Baldwin & Clark, 1997: 84). The production of a single automobile product involves steps including producing and procuring a wide range of parts, as well as final assembly. Modular production reduces the complexity of final production by creating larger composite units called 'modules' through the sub-assembly of multiple components that are then put into the final assembly line, unlike the conventional method where individual components are procured and then assembled.

In the automobile industry, modular production has been led by German automakers, and spread to the United States through Europe. After encountering limitations with the Japanese-style lean production system, from the 1990s Hyundai Motor tried modular production as an alternative production technology system to catch up with Japanese automakers. Modular production has even been referred to as another revolution in production system that is comparable to lean production (Helper et al., 1999). It is a strategy for finished carmakers to reduce product development and production costs and dramatically increase production flexibility and quality by ensuring that specific first-tier component manufacturers are fully responsible for the design, production, and delivery of modular parts (Hyundai Mobis, 2007: 541–542; McAlinden et al., 1999).

Hyundai Motor began to adopt modular production on a large scale in the 2000s. This was part of the Hyundai Motor Group's aggressive management strategy to develop into one of the world's top five automakers. As part of this drive, a modular production task-force was formed in the early 2000s to promote the modular production strategy. A specialized parts supplier called 'Hyundai Mobis' was established as an affiliate in 2000 based on the existing Hyundai Precision Industries' auto parts division. Hyundai Mobis later grew into a core modular parts producer that drives modular production at Hyundai Motor and Kia. Beginning with the modularization of chassis parts, electric devices and components, and trim assembly processes, Hyundai Mobis has continued to develop and supply new modular parts and expanded the scope of modular production by increasing the proportion of modular parts whenever a new model is introduced.

At the top of the modular production strategy is a platform sharing strategy that reduces production costs and product development costs by designing multiple models to have the same basic structure. After integrating with Kia, Hyundai Motor established a strategy of modular production based on the foundation of platform sharing. Modular production was able to not only shorten the product development period but also significantly increase production flexibility through the sharing of parts (Interviews with a Hyundai Motor executive in 2012).

Table 4.2 shows the modularization for each vehicle type in the mid-2000s. Hyundai Motor Group selected six modular parts as official modularization targets; the front chassis, rear chassis, front-end, cockpit, roof (headlining), and door shield (Hyundai Motor Company, 2005). As shown in this figure, the modularization of these six parts through affiliates such as Hyundai Mobis was almost complete by the mid-to-late 2000s (Interviews with a Hyundai Motor executive in 2012). As a result, Hyundai Motor is believed to have achieved a modular production rate comparable to that of German automakers such as Volkswagen in the late 2000s.

Modular production makes it possible to reduce the number of shop-floor workers and thus reduce labor costs by facilitating automation and standardization in automobile assembly. In addition, since modular production involves the pre-assembly of many complex parts before putting them into the final assembly line, a large chunk of the process is transferred to modular parts suppliers. This functional transfer extends not only to simple processing and sub-assembly of parts, but also to development, quality, and inventory responsibility for modular parts.

Modular production also provides an excuse to outsource other parts that are unrelated to modular parts. In fact, in addition to typical modular parts such as cockpit modules, front-end modules (FEM), and chassis modules, many other parts were packaged as 'modular production' by simply outsourcing or changing delivery units in Hyundai Motor's production system. As a large part of assembly and production was outsourced in this way, these functions were scaled back and simplified within Hyundai Motor. This modular production strategy is closely linked to the characteristics of Hyundai Motor's production system, which developed under the conditions of engineer-led automation and informatization, and confrontational labor–management relations. In other words, Hyundai Motor's production system demonstrated a certain selective affinity for modular production.

The characteristics of Hyundai Motor's production system specifically promoted the progress of modular production in three ways. First, Hyundai Motor's automation and information technology capabilities greatly contributed to advancing the outsourcing of major components and modular production. As a latecomer, Hyundai Motor was able to

Table 4.2 Hyundai Motor's modularization

Category	*Entry Car*		*Compact Car*				*Mid-size Car*		*SUV*				*MPV*
	Click	*MC*	*XD*	*HD*	*Tuscani*	*Lavita*	*TG*	*NF*	*JM*	*(EN)*	*CM*	*Trajet*	*(TG)*
Cockpit Module				●	●		●	●		●	●		●
Main Crashpad	●	●	●	●	●	●	●	●	●	●	●	●	●
Cowl Cross Member	●	●	●	●	●	●	●	●	●	●	●	●	●
Crashpad Wiring	●	●	●	●	●	●	●	●	●	●	●	●	●
Steering Column		●		●	●		●	●	●	●	●	●	●
Heater/Blower/Eva		●		●	●	●	●	●		●	●		●
Loop Module		●		●				●	●	●	●		●
Headlining	●	●	●	●	●	●	●	●	●	●	●		●
Loop Wiring	●	●	●	●	●	●	●	●	●	●	●		●
Lamp		●		●					●	●	●		●
Assist Handle		●		●			●	●	●	●	●		●
Door Shield Module	○	○	○	○	○	○	○	○	○	○	○	○	○
Door Inner Module			●	●			●	●	●	●	●	●	
Door Wiring				●			●	●	●	●	●		
Door Trim Seals				●			●	●	●	●	●		
Front End Module	●	●		●			●	●	●	●	●		●
Cooling Module	●	●	●	●	●	●	●	●	●	●	●	●	●
Radiator Member	●	●		●			●	●	●	●	●		●
Headlamp	●	●		●			●	●	●	●	●		●
Front Bumper Back Beam	●	●		●			●	●	●	●	●		●
Chassis Module	○	○		○					○	○	○		
Front Chassis Module	○	○	○	○	○	○	●	●		○	○	○	○
Engine/Transmission	○	○	○	○	○	○	○	○	○	○	○	○	○
Front Crossmember	●	●	●	●	●	●	●	●	●	●	●	●	●
Front Strut	○	●	●	●	●	●	●	●		●	●	●	●
Rear Chassis Module	○	●	●	●	●	●	●	●	●	●	●	●	●
Rear Crossmember		●	●	●	●	●	●	●	●	●	●	●	●
Rear Strut			●	●	●	●	●	●	●	●	●		

Note Black circles refer to modularized parts, while white circles refer to partially completed parts
Source Author's own adaptation from Hyundai Motor's internal documents in 2006

avoid repeating the trial and error process of more advanced companies, which was an advantage in introducing the latest facilities and programs to product development and production technology. Hyundai Motor's engineers actively sought to introduce and absorb advanced technologies within an organizational culture that did not hold individuals accountable even if they failed. Hyundai Motor's technological capabilities also contributed to the successful promotion of modular production by making it easier to manage the parts companies involved in modular production (Interviews with a Hyundai Motor executive in 2012).

Second, Hyundai Motor did not expect shop-floor workers to be actively involved in the process due to the history of confrontational labor–management relations, and thus adjusted its production system in a way that reduces the level of dependence on shop-floor workers' skills. This made it easier to promote modular production without compromising product quality. Hyundai Motor has been able to outsource and modularize a large part of the company's final assembly line because it reduced reliance on in-house skilled workers by promoting automation and informatization. In addition, there was no difficulty in maintaining productivity and quality even when outsourced modular parts were produced and delivered by low-skilled workers from parts suppliers (Interviews with a Hyundai Motor executive in 2012).

Finally, modular production allowed Hyundai Motor to achieve numerical flexibility through the modular parts supplier Hyundai Mobis. The work organization of Hyundai Motor's production system, which has the characteristic of increasing the automation rate of the production process and minimizing dependence on the skills of shop-floor workers, can realize its potential through numerical flexibility. Hyundai Motor attempted to follow a strategy of outsourcing and hiring workers in less stable work arrangements in the 2000s, but this ran into inevitable opposition from labor unions. Modular production made it easy for Hyundai Motor to transfer a large chunk of the final assembly process to Hyundai Mobis in the form of modular part sub-assembly. Hyundai Mobis hired 'non-regular' workers through in-house subcontracting to direct produce the modular parts. The company minimized the need for worker skills by applying the latest process technology to exploit non-regular workers. As a result, Hyundai Mobis was able to easily replace a significant portion of full-time employees at Hyundai Motor with non-regular workers.

4.3.2 *Flexible Automation and Informatization*

In a situation where Hyundai Motor is able to save on shop-floor skills, the company developed flexible production by utilizing the potential of technology rather than labor. After renovating the production line, a mix of models are now produced on a single line in the final assembly shop. In other words, the company deliberately created general-purpose facilities that can produce different models with minimal equipment remodeling, programming, and tool replacement, with enough space to deliver multiple models in sequence. As the number of production units per model decreases due to the recent transition toward future cars and the segmentation of market demand, mixed production is expected to grow even further.

The flexibility of the production line has been further strengthened in combination with progress in modular production. Sub-assembling large and bulky parts in module units before putting them into the final assembly line in sequence saves on space for the assembly of parts, which generally increases with multi-model production, thereby enhancing productivity as well (Interviews with Hyundai Motor, 2011).

Hyundai Motor's flexible automation has been supported by the information system. The development of Hyundai Motor's information system came about due to the continued conflict in labor–management relations. In a situation where the company could not expect active participation from shop-floor workers amid distrust between labor and management, Hyundai Motor's information system was developed to be technology-oriented and skill-saving in a way that did not depend on the skills of shop-floor workers. In other words, it minimizes the interface between humans and the information system.

The informatization that supports Hyundai Motor's flexible production was completed through the Enterprise Resource Planning (ERP) system introduced in 2009. As shown in Fig. 4.1, ERP is an information system that integrates, analyzes, and manages information across all areas of business activity. From September 2009 to August 2010, Hyundai Motor rolled out the ERP system to connect and integrate information systems developed in individual arenas such as MRP. Purchased from world-leading software company SAP for a total cost of KRW 32.8 billion, ERP includes BOM, MRP, VAATZ (Value Advanced Automotive Trade Zone for purchasing of parts), VMI (Vendor Management Inventory for delivery orders), ALC, and more. This enables the company to not

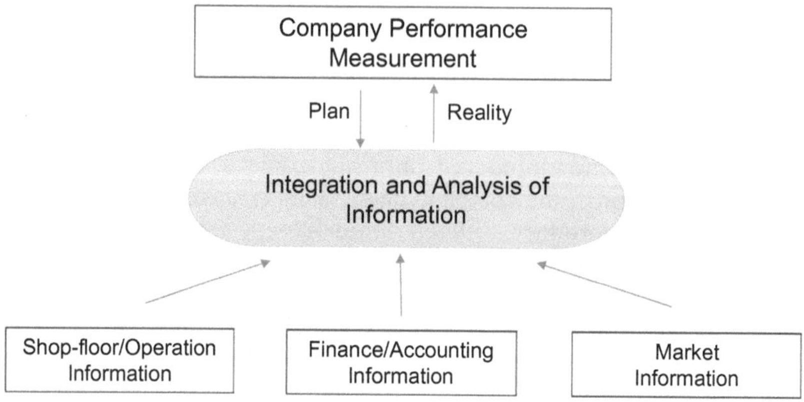

Fig. 4.1 Map of ERP concept (*Source* Author's own creation)

only share information in real time and increase transparency by linking existing systems, but also increases efficiency in production management by facilitating prompt response to changes (Jo & Cho, 2012).

What is noteworthy here is that Hyundai Motor was the first among major global automakers to fully adopt the ERP system. This huge investment decision can be regarded as the result of a combination of technologically oriented problem-solving efforts by engineers and a strong motivation on the part of executives to bypass confrontational labor–management relations and minimize the importance of shop-floor worker skills.

4.3.3 *Skill-Saving Work Organization*

Hyundai Motor's skill-saving work organization is based on teams consisting of around 30 workers, with each team made up of 3–4 groups of 7–8 workers. Each team consists of a foreman, general workers, and staff who fill in worker vacancies. The field leader is in charge of overall operations and management, such as allocating work and being responsible for product quality. In most teams, there is a voluntary job rotation. However, since this is mostly aimed at reducing the monotony of the work, it is difficult to argue that this represents a systematic multi-skilled labor program at the company level (Jo et al., 2008). Systematic skill formation is largely absent for workers on the shop-floor. In this respect,

Hyundai Motor's skill formation system contrasts with that of Toyota, where all employees are classified as simple, highly versatile, or specialized workers and the company strives to systematically improve their skills (Koike & Nakama, 2001).

> Most workers have been here for more than 20 years and always worked in the same department, so in terms of experience they are multi-skilled in that they can perform work across multiple processes. However, since only a small minority have a strong motivation to develop their job skills, most workers at Hyundai Motor perform simple, repetitive tasks that they can master in a short period of time. The company has virtually given up on systematically improving their skills. (Interviews with a site supervisor of Hyundai Motor in 2013)

Hyundai Motor has developed a fool proof system so that quality is not affected by the level of workers' skills. This operates through three steps. First, workers scan the barcode of the component while looking at work instructions on the monitor, check the okay mark, and install the component, which prevents other component from being used or mounted. Second, when assembling parts, the system automatically checks whether bolts or nuts are tightened to within a certain value. If the value is exceeded, the monitor displays this in red to prevent incorrect assembly. Third, the quality of assembled parts is automatically checked in the final process in order to make up for the possibility of mistakes in the visual inspection. The combination of these systems makes product defects rare. However, applying advanced technology that minimizes defects caused by human error also implies that the company's expectations for skill upgrades among shop-floor workers are low.

Hyundai Motor's disregard for skill formation has made the job division between 'regular' and 'non-regular' workers meaningless, serving as a technical condition that has encouraged increasing numbers of non-regular workers. At Hyundai Motor's production sites, there is almost no difference in skill level between regular and non-regular workers (Interviews with a site supervisor of Hyundai Motor in 2013).

Figure 4.2 shows the typical work organization of Hyundai Motor's final assembly shop. In 2012, the day shift group (group A) assembled 30 units per hour with a mixed assembly of Starex and Genesis Coupes. As mentioned earlier, each team consists of three groups of 7–8 workers for a total of around 30 members. This includes up to three foremen who work

		Final			Final(F0H)												
		1	2	3	34	35	36	37	38	39	40	41	42	43	44	45	46
Group A	LH	0	0	0	☆	00	0	☆	☆	0	0	★	0	★	00	0	0
	RH	0	0		0	0	0	☆	☆ ☆	0		★	★	★★	★	0	00
Group B	LH	0	0	0	★	★0	0	★	☆	0	0	★	0	★	★0	0	0
	RH	0	0		★★	★	0	★	☆ ☆	0		★	★	★★	0	0	00

		OK Fix									Water Leaking	Side
		1	2	3	4	5	6	7	8	9		
Group A	LH	0	0	★	★	★	★	0	0	0	000	000
	RH	0	0	★	★	★	★	0	00	0	00	00
Group B	LH	0	0	★	★	0	0	0	0	0	000	000
	RH	0	0	★	★	0	0	0	00	0	00	00

Fig. 4.2 Final assembly line layout of Hyundai Motor's Ulsan plant (as of August 14, 2012) (*Note* 0 indicates direct employment, while ★☆ indicates worker from subcontractor. A difference in color means workers belonging to different subcontractors; *Source* Author's own adaptation from Hyundai Motor's internal documents in 2012)

in place of staff to fill in for absences. In the day shift team, 1–2 workers are assigned to either the left or right side of each process from process 34 to process 46 depending on the workload. What is noteworthy is that 17 regular full-time employees and 13 non-regular subcontract workers work in the same team. Non-regular subcontract workers have been assigned every time the shop changes the model in production, usually for handling heavy items or difficult processes that regular workers avoid.[1] This means the job rotation is more about which tasks are commonly avoided than job variety, and the system is not working as planned.

[1] Hyundai Motor's use of in-house subcontract workers sparked severe controversy that the company violated the law prohibiting the direct use and management of dispatched workers on the shop-floor. In fact, since 2010, Hyundai Motor has been ruled by the Supreme Court one after another that the direct use and management of in-house subcontract workers is illegal. Following the court's ruling that the company should convert them to regular or full-time workers, Hyundai Motor drastically reduced the size of in-house subcontract workers by hiring about 6000 new in-house subcontract workers as regular or full-time in 2016. As a result, contract workers, instead of in-house subcontract workers, are now assigned and work in the same group with regular or full-time workers.

4.3.4 *Multi-Tiered Quality Control*

Hyundai Motor's quality control system reveals the characteristics of the company's production system, which does not rely heavily on the skills of shop-floor workers. Quality control at Hyundai Motor is carried out in multiple layers. The lack of shop-floor skills is compensated for in multiple layers through pre-quality control via trial production, quality control for outsourced parts, and an additional quality controls after production is completed.

First, Hyundai Motor attempts to identify and solve all problems that could occur when starting mass production of a specific model through pilot production, as discussed in more detail in Chapter 3. When the company develops a new model, more than 100 vehicles are tested for 1.5 months each in two stages (P1 and P2) at the pilot plant located in Namyang Research and Development Center. This allows Hyundai Motor to discover problems and modify the design in advance (Interviews with a Hyundai Motor executive in 2012). The new car development process is markedly different from Toyota's. Toyota performs small-scale test production using separate facilities at the P1 stage, but the majority of the test production is subsequently carried out on the mass production line. Toyota is highly dependent on the capabilities of shop-floor workers to solve problems and put forward suggestions (Oh, 2018). Hyundai Motor also undergoes a process of adjusting problems that could appear during mass production by involving skilled workers in both the P1 and P2 stages. However, engineers are mainly responsible for actually implementing most of the problem-solving and design changes.

Second, Hyundai Motor has improved parts quality management. The Quality Headquarters directly inspects parts instead of entrusting the initial quality to parts suppliers. In addition, problems that occur during the trial production process are addressed and solved through design changes by providing feedback to parts suppliers. Through this system, Hyundai Motor has striven to improve not only the initial quality of finished vehicles but also durability. The progress of modularization allows modular component suppliers to take responsibility for the interim and sub-assembly of components, so component quality continues to improve.

Finally, Hyundai Motor has also established a triple quality inspection procedure after the production of finished vehicles is complete. For export products, quality defects are discovered and resolved through PDI

(Pre-Delivery Inspection) at the shipping pier, an inspection at the local unloading pier, and a dealer inspection. The cause of any defect found is identified, and the responsible department head is held accountable. It is true that this multi-layered quality inspection entails additional costs. However, considering the recall cost and the brand value deterioration caused by poor quality, there appears to be no other option (Interviews with a Hyundai Motor in 2008).

4.3.5 *Skill-Saving Work Organization, Labor–Management Relations, and Labor Market Segmentation*

Production systems develop through interactions with labor relations. In this section, the skill-saving work organization of Hyundai Motor's production system can be explained in connection to labor–management relations. What is the relationship between flexible production and labor relations? In general, involvement and participation from shop-floor workers is viewed as necessary for flexible production. For example, at Toyota, workers actively participated in a workplace improvement activity called 'Kaizen' as the union complied with the company's policy. This has been regarded as one of the key elements of the company's lean production system. Volkswagen implements another form of labor participation in that workers are involved in joint decisions through the 'work council' (Jung et al., 2008).

However, worker participation is not the only thing that enables flexible production. The development of Hyundai Motor's production system is connected to the company's hostile labor–management relations. Hyundai Motor's production system was characterized by a focus on engineer-centered technology and greater emphasis on oppressive control than participation among shop-floor workers, which served as a basis for producing confrontational labor–management relations by creating distrust in the company. At the same time, hostile labor–management relations made it difficult for workers to actively participate, further incentivizing Hyundai Motor to develop its production system in a technology-oriented direction.

Prior to 1987, Hyundai Motor mass-produced its own standardized products and labor unions did not exist. Under the support of Korean developmental state, management exercised strong unilateral power over workers. Hyundai Motor's intensive problem-solving was accompanied by coercive labor controls for shop-floor workers in a militant fashion.

> At that time, you weren't allowed to have your ears covered by your hair. Now, I grow my hair long and have my own style, but back then I couldn't dye my hair or let it grow long enough to cover my ears. If they thought you were having trouble hearing, the security guard would pull out a pair of scissors and cut your hair himself. There was a barber shop inside the yard, and you didn't go there, you couldn't go to work that day. (...) Simply put, it was the same as the government's crackdown on men with long hair in the 1970s. And if you did something wrong, you'll get kicked in the shin. (...) So hit like that. (Interviews with a shop-floor worker of Hyundai Motor in 2015)

In 1987, labor disputes against oppressive labor controls broke out across Korea. 'Democratic' labor unions were established in numerous workplaces across the country as part of a wave of large-scale labor disputes, referred to as the 'Great Workers' Struggle.' Oppressive labor controls and poor working conditions emerged as important issues in labor–management relations.

Since the establishment of Hyundai Motor's labor union in 1987, confrontational relations between labor and management have continued over excessive labor controls and poor working conditions. As the unilateral top-down approach of the past is no longer feasible, the company has now begun to push for automation in a way that reduces the importance of workers. Despite the higher investment cost, it was more efficient for Hyundai Motor to increase the rate of automation to improve productivity and quality instead of encouraging active involvement and participation from workers. The old labor-intensive production system created resistance and confrontational labor–management relations, and this served as a catalyst for the company to promote skill-saving work organization.

Skill-saving work organization is characterized by numerical flexibility that relies on low wages and flexible employment, as opposed to functional flexibility that utilizes workers' skills. In fact, Hyundai Motor has set up the technical conditions to achieve numerical flexibility by maximizing the separation between conception and execution and minimizing dependence on shop-floor skills so that tasks can easily be performed even by unskilled workers.

Of course, there are also some factors that limit the effectiveness of numerical flexibility. The relationship with the labor union is one such factor. Since numerical flexibility is directly connected to job insecurity, labor unions and workers have strongly opposed this process. As a result, measures that restrict numerical flexibility have been introduced through collective agreements, even though their impact is limited.[2] In this regard, rapidly pushing to implement numerical flexibility could lead to strong opposition from the labor union and destabilize production.

However, opposition from the labor union and instability in labor–management relations were also major contributors to the original decision to pursue numerical flexibility. In order to reduce the labor union's influence on strategic management activities and minimize the impact of unstable labor–management relations on stable profit-seeking, Hyundai Motor has shifted its production functions to areas where it is difficult to organize labor unions (outsourcing), and expanded the use of non-regular workers.

This was supported by the 'employment stability agreement' signed between labor and management. In June 2000, both sides signed a full-employment security agreement. The application of this document was limited to regular workers provided by the labor union, and it excluded in-house subcontracted non-regular workers under the company's *de facto* control. Since then, Hyundai Motor has expanded outsourcing and the use of non-regular workers to reduce costs and pursue numerical flexibility while maintaining the employment status and wages of regular workers. As modular production developed, so-called 'non-regular shops' also emerged from Hyundai Mobis, and this practice soon spread to other companies. When Hyundai Motor outsources a large part of final assembly to modular parts suppliers by leveraging modular production, new modular parts providers assign all modular parts to be sub-assembled by non-regular workers. This strategy is aimed at reducing costs and improving numerical flexibility by maximizing the use of non-regular workers. At the same time, it can be viewed as a strategy to block union interference in the production process.

In this way, the technical conditions for numerical flexibility resulting from Hyundai Motor's skill-saving work organization have led to

[2] For example, through a collective agreement with the company, the Hyundai Motor Branch of the Korean Metal Workers' Union stipulates that the union must be given notice 60 days in advance of any plan affecting employment, such as outsourcing including modular parts sub-assembly, subcontracting, or task conversion, and a joint labor–management committee must be formed to deliberate on and resolve the issue (Metal Workers Union Hyundai Motor Branch 2009 Collective Agreement, Article 40 Paragraph 1).

outsourcing and reduced job security. The shift toward modular production at Hyundai Motor in the 2000s served as an excuse to outsource a significant part of mass production. As modular production developed, the subcontracting structure became stratified, and the segmentation of the labor market deepened in the hierarchy between contractors and subcontractors, and first-tier subcontractors and second-tier subcontractors (see Chapter 5).

The increased use of non-regular workers, which allows Hyundai Motor to flexibly adjust the workforce as needed in a skill-saving work organization, has led to a segmented structure between regular and non-regular workers in the firm internal labor market. The labor–management relationship limits outsourcing and non-regular workers to a certain level, but also served as a catalyst that encouraged such practices to circumvent the union in the first place. As a result, not only has the labor market become segmented between companies, but a divide has emerged between regular and non-regular workers within the company.

4.3.6 *Comparison with Toyota Production System*

Hyundai Motor has greatly improved the flexibility of production technology throughout the production process. The company has not only increased the fidelity of product design, but also developed a bill of materials (BOM) for component specifications. Communication between hardware such as robots and the assembly line control (ALC) system is seamless. In the welding shop, the feeding of car body parts and the angle adjustment of welding robot have been performed accurately. At the painting shop, color selection, paint quality, internal air management, grouping colors together, and decisions on input order have been greatly improved. In the final assembly shop, the just-in-sequence supply of parts is carried out smoothly. Hyundai Motor's rapid growth seems to be largely due to the development of this production technology (Interviews with a Hyundai Motor executive in 2006). It is worth noting that Hyundai Motor's exclusion of shop-floor workers from flexible automation is very different from Toyota's lean production system, which leverages the intellectual ability of workers as a key element.

In Toyota's production system, workers not only become multi-skilled through job rotation and cycling, but are also trained into skilled workers with a high level of problem-solving ability and the capacity to respond

Table 4.3 A comparison of Toyota's production system and Hyundai Motor's

	Toyota's production system	*Hyundai Motor's production system*
Production technology	Labor-participatory flexible automation	Engineer-led and labor-excluding flexible automation
Work organization	Job rotation, multi-skilled workers	Skill-saving
Organizational capability	Problem-solving centered on the shop-floor	Multi-layered management centered around engineers and middle managers

Source Author's own creation

to changes and irregularities through a combination of OJT and Off-JT (Oh, 2018). In contrast, most shop-floor workers lack systematic skill formation in Hyundai Motor's system, although some indirect departments such as QC and maintenance do play an important role. As a result, it is mainly engineers who have been directly responsible for tasks that require a high degree of skill.

While workers on the shop-floor are at the heart of problem-solving in Toyota's production system, in Hyundai Motor's system the organizational capabilities of production processes compensate for the rigidity of a work structure that lacks flexibility. One example of this is the informatization of organizational capabilities led by engineers and multi-layered management in the quality management system. Table 4.3 compares Toyota's production system and Hyundai Motor's production system.

Studies of production systems in the automobile industry have mainly conceptualized Hyundai Motor's system as a 'peripheral' or 'marginal' production system compared to that of automobile manufacturers in countries with competitive advantages such as Fordism or Toyotism. However, this categorization only emphasizes Hyundai Motor's inferiority as a latecomer, and fails to take into account many of the system's unique characteristics.

4.4 Conclusion

Hyundai Motor's production system is characterized by flexible automation and skill-saving work organization. While pursuing flexibility in production by utilizing automation technology, the company simplified and standardized the labor process as much as possible to minimize the need for skilled workers on the shop-floor. This is different from the Toyota's production system, which has a flexible work organization that maximizes the intellectual ability of workers as a key factor.

Skill-saving work organization developed through interaction with Hyundai Motor's unique labor–management relations. Oppressive labor controls gave rise to distrust on the part of labor and hostile labor relations within the company, and this served as a catalyst for pursuing skill-saving work organization that largely excludes participation from labor. In a situation where shop-floor workers are not expected to participate, a dual and triple quality management system using engineers and indirect departments compensates for the lack of skills of shop-floor workers. However, the excessive workload for engineers could lower their morale as middle managers and have the unwanted side effect of making them avoid the shop-floor.

In this skill-saving work organization, labor is either replaced by machines or becomes a target of numerical flexibility. The proportion of non-regular workers has also expanded, and a segmentation between existing regular workers and non-regular workers has occurred in the firm internal labor market, with the gap widening over time. This segmentation is another characteristic of Hyundai Motor's production system.

One interesting fact is that the higher the proportion of overseas local production, the greater the competitive advantage of Hyundai Motor's production system. Since flexible production is achieved through automation and informatization and there is a low degree of dependence on worker skills, it is relatively easy for Hyundai Motor to transfer the system overseas. The company's overseas plants in China, the United States, and Europe have surpassed the level of Korean plants in productivity and quality just two to three years after beginning mass production, which stems from the characteristics of the system itself. In contrast, Toyota has not been able to surpass the performance of its domestic plants even decades after launching local production overseas. In this sense, Hyundai Motor's production system shows a clear competitive advantage in overseas production.

References

Baldwin, C. Y., & Clark, K. B. (1997). Managing in the Age of Modularity. *Harvard Business Review, 75*(5), 84–93.

Chung, M.-K. (1997). Introduction of New Production Methods and Changes in Human Resource Management: Focusing on the Case of Hyundai Motors Asan Plant. *Korean Journal of Labor Studies, 3*(1), 81–108. (In Korean).

Helper, S. R., MacDuffie, J. P., Sako, M., Takeishin, A., & Warburton, M. (1999). *Modularization and Outsourcing: Implications for the Automotive Assembly* (IMVP Working Paper). http://imvp.mit.edu/papers

Hyundai Mobis. (2007). *30 Years of Hyundai Mobis*. Hyundai Mobis Company. (In Korean).

Hyundai Motor Company. (1987). *20 Years of Hyundai Motor Company: 1967–1987*. Hyundai Motor Company. (In Korean).

Hyundai Motor Company. (2005). *Modularization Direction*. Hyundai Motor Company. (In Korean).

Jo, H. J., & Cho, J. E. (2012). Does Hyundai Motor Company's Production Management Converge to 'Pull' Production? A Study on the Evolution of Demand-driven Production Management through Information System. *Korean Journal of Sociology, 46*(5), 233–257. (In Korean).

Jo, H. J., & Lee, B.-H. (2008). A Study on the Historical Evolution of Hyundai Production System: Examining the Adoption of Japanese Production System. *Journal of Korean Social Trend and Perspective, 73*, 231–264. (In Korean).

Jo, H. J., Park, T.-J., Cho, S.-J., Lee, J. H., & Paik, S. Y. (2008). *An Exploratory Study on Hyundai Production System*. Hyundai Motor Company. (In Korean).

Jung, S. G., Jo, H. J., Lee, S., & Kim, A. (2008). *Skill Formation and Wage System*. Korea Research Institute for Vocational Education & Training. (In Korean).

Kim, Chulsik. (2011). *The Growth of Large Firm and the Destabilization of Labor: An Analysis on the GVC, Production System and Employment Relationship in Korean Automobile Industry*. Baeksanseodang. (In Korean).

Koike, K., & Nakama, H. (2001). *Manufacturing Skills*. Keizai Inc. (In Japanese).

Kuhlmann, M., & Schumann, M. (1997). Patterns of Work Organization in the German Automobile Industry. In K. Shimokawa, U. Jürgens, & T. Fujimoto (Eds.), *Transforming Automobile Industry: Experience in Automation and Work Organization* (pp. 289–304). Springer.

McAlinden, S. P., Smith, B. C., & Swiecki, B. F. (1999). *The Future of Modular Automotive Systems: Where are the Economic Efficiencies in the Modular-Assembly Concept?* (UMTRI Report No. 2000-24-1).

Oh, J. (2018). A Study on the Innovation of Production and Development System in the Japanese Automobile Firms. *The Korean-Japanese Journal of Economics and Management Studies, 79*, 115–140. (In Korean).

Piore, M. J., & Sabel, C. F. (1984). *The Second Industrial Divide: Possibilities for Prosperity*. Basic Books.
Womack, J. P., Jones, D. T., & Roos, D. (1990). *The Machine that Changed the World*. HarperPerennial.

CHAPTER 5

Supplier Relations—Extended Quasi-Vertical Integration

5.1 Introduction

In this chapter, we analyze the characteristics of Hyundai Motor's parts supply structure as aspects of the external processes of the company's agile production system. Hyundai Motor has a supply chain that is markedly different from those of other automakers, which we conceptualize as 'extended quasi-vertical integration.' We argue that the main factors behind Hyundai Motor's rapid growth in the 2000s were an extended network of vertically quasi-integrated inter-firm relationships, combined with engineer-led innovation and skill-saving work organization.

MacDuffie (2013) was the first to conceptualize Hyundai Motor's parts supplier relationships as quasi-vertical integration in his work on the modularization of the global automotive industry. According to MacDuffie, Hyundai Motor adopts advanced modularization in production, while Hyundai Mobis, a key affiliate in the Hyundai Motor Group, has gradually evolved from a simple provider of modular parts into a key supplier that is responsible for designing modular parts. Rather than independently taking charge of design work, Hyundai Mobis performs modular part design in a closely interdependent relationship with Hyundai Motor. The relationship between the two entities is not a trading relationship between independent companies, but a closely interdependent relationship between affiliates that belong to the same group, which can be called quasi-vertical integration.

H. J. Jo et al., *Agile Against Lean*,
https://doi.org/10.1007/978-981-99-2042-6_5

The relationship between Hyundai Motor and Hyundai Mobis that MacDuffie (2013) labels quasi-vertical integration is an affiliate relationship within a *chaebol* group, a network that is closely linked to equity ownership. However, the supply chain is not limited to relationships between affiliates within the group. Each individual affiliate is interconnected not only with each other, but also with non-affiliates and external entities to form a 'network with networks' (Dicken, 2015: 130). Accordingly, the nature of the supply chain cannot be understood solely by looking at the group's internal network.

We conceptualize Hyundai Motor's supply chain as 'extended quasi-vertical integration' because it represents a closely linked network that resembles vertical integration within the firm combined with a network of legally independent firms, irrespective of whether they are affiliated or not. However, contrary to the assumption of MacDuffie (2013), this extended quasi-vertical integration includes not only legally separate affiliates within the *chaebol* group, such as Hyundai Mobis, but also legally independent and non-affiliated suppliers from outside the group. A vertical hierarchy similar to the hierarchy within the firm also exists in the external network through relationships between non-affiliates outside the group.

One important catalyst that led to the current form of Hyundai Motor's extended quasi-vertical integration network was the modular production that took place in the 2000s. However, the prototype of this network can be traced back to the early days of Hyundai Motor's founding in the mid-1970s. Accordingly, Sect. 5.2 examines the formation process of Hyundai Motor's extended quasi-vertical integration network before the Asian financial crisis at the end of 1997. Section 5.3 then analyzes how extended quasi-vertical integration became systematized through modular production in the 2000s. In Sect. 5.4, we present the factors that enable Hyundai Motor's seemingly contradictory quasi-vertical integration to be maintained. Section 5.5 compare the experience of Hyundai Motor with the relationship between Japanese automakers such as Toyota and their parts suppliers. In the final section, we review the implications and limitations of Hyundai Motor's extended quasi-vertical integration model.

5.2 Formation of Extended Quasi-Vertical Integration—Before the 1997 Financial Crisis

5.2.1 *The Formation of the Subcontracting System*

After the Korean War, the Korean automobile industry started out as a number of companies that refurbished or assembled used cars based on old military vehicles, scrapped vehicles, and used parts. In the 1960s, disassembled parts from automobile companies in developed countries were assembled into so-called 'knocked-down' vehicles. The industry first reached the stage of developing its own model passenger cars in the 1970s. In the late 1970s, Hyundai Motor began to pursue an export strategy and the company established a mass production system and started exporting to the United States in the 1980s.

During this period of rapid development, the Korean automobile industry established a wide range of industrial links through the procurement of parts. The development of Korean parts makers was largely driven by the industrial policy of the developmental state. In May 1974, the Korean government established the Long-Term Automobile Industry Promotion Plan with the goals of developing domestic automobile models and improving the localization rate for parts, separating finished vehicle assembly from parts and components plants, and promoting a policy of specialization whereby each manufacturer would focus on one part or component. As a result, parts makers began to separate from automakers. Subsequently, the Small and Medium Business Subcontracting Promotion Act enacted in 1975 (and slightly revised in 1978) introduced the 'designated subcontracting' scheme. Under this act, if the government designated a part for local specialization, it was mandatory to form a subcontracting relationship between the parent company and the parts maker. Government tax relief and financial support were provided to contractors and subcontractors that formed a subcontracting relationship.

In a major amendment to the Small and Medium Business Vertical Integration Promotion Act in 1982, policy shifted toward promoting private-led subcontracting systems, making it mandatory for each automaker to form a suppliers' council. As a result, each automaker formed their own suppliers' council starting in 1984, and first-tier suppliers then formed their own council of suppliers. As of December

1991, there were a total of 17 suppliers' councils in automakers and first-tier suppliers as parent companies (see Table 5.1).

Hyundai Motor formed the 'Hyundai Cooperative' as a suppliers' council in 1984, which had 238 parts makers as active members as of the end of 1991. In particular, compared to the suppliers' councils of other automakers, the Hyundai Cooperative had strong leadership from its parent company. Only companies with a high degree of dependence on Hyundai Motor could become members due to the strict membership qualifications that required more than two years of transactions with Hyundai Motor and more than 40% of the supplier's total sales. In addition, Hyundai Motor covered almost all the expenses of the Cooperative's meetings, and the secretariat office established within Hyundai Motor was in charge of all operations including planning, execution, and accounting. In this regard, the Cooperative can be said to be more akin to a management council of Hyundai Motor's parts makers than an independent council (Chung, 1994: 31).

Table 5.1 Suppliers' council membership by parent company (as of 1991)

Name of parent company	*Date of establishment*	*No. of member companies*
Kia Motors	1977.11.18	168
Daewoo Motors	1984.03.15	182
Hyundai Motor	1984.04.12	238
Ssangyong Motors	1984.03.30	38
Asia Motors	1985.10.11	149
Daewoo Heavy Industry	1984.03.30	188
Hyundai Motor Services	1984.04.12	40
Daelim Motors	1984.11.29	97
Hyosung Machinery Industry	1985.11.22	95
Kia Machinery Industry	1985.12.26	78
Korea Spicer Industry	1986.12.20	36
Seil Heavy Industry	1987.06.01	37
Mando Machinery	1987.12.28	74
KIA Precision Machinery	1988.03.29	57
Daewoo Precision Industry	1987.12.02	24
Poongseong Electric	1989.12.20	27
Daewoo Components Industry	1990.05.09	27

Source Korea Auto Industries Coop. Association (KAICA). 1992. Automobile Industry Handbook. p. 84 and author's own adaptation

Under the strong leadership of Hyundai Motor, the members of the Hyundai Cooperative formed a captive business relationship with Hyundai Motor. With the launch of the council of suppliers, a vertical subcontracting system was established in which Hyundai Motor had strong authority to manage both affiliated and non-affiliated parts makers.

5.2.2 *Captive Relationships and Paternalistic Control*

With the formation of the Hyundai Cooperative, Hyundai Motor took the lead in supporting parts makers. By establishing the subcontracting system, Hyundai Motor provided stable demand to parts makers to ensure that parts makers belonging to the Cooperative did not go bankrupt. In addition, Hyundai Motor supported the stability and growth of parts makers by providing technical and management guidance and financial support (Jo & Kim, 2013).

First, Hyundai Motor arranged parts makers to form technology partnerships with advanced overseas parts makers and receive technical training. For example, Hanil-Ewha, the chair company of the Cooperative at the time, was able to import door trim and headlining technology from Japan's Ikeda Corporation, Kasai Industries, and Tenryu Industries through an arrangement made by Hyundai Motor.

Second, Hyundai Motor helped parts suppliers enhance their productivity by improving plant layouts and processes. For example, distance between processes was reduced by rearranging equipment into a straight line. This made processes more efficient and uniform, thereby reducing the inventory and staff required.

Third, Hyundai Motor focused on preventing quality problems at parts makers by conducting intensive guidance on quality assurance. Guidance from overseas experts was carried out regularly, including visits from Hyundai Motor's Japanese senior adviser Mr. Arai.

Finally, Hyundai Motor pushed to eliminate bottlenecks and reduce defects by providing technical guidance on automation and labor saving. This was intended to improve productivity, enhance quality, and reduce costs. In addition, Hyundai Motor helped parts makers develop technological capabilities in fields such as joint purchasing of raw materials and support for new product development (Hyundai Motor Company, 1987: 477–480, 607–608).

This support further entrenched the captive business relationships with the parent company. As mentioned above, it was necessary to maintain a

Table 5.2 Number of parent companies for first-tier parts suppliers (as of 1997)

No. of parent companies	*1*	*2*	*3+*	*Total*
No. of Suppliers (%)	619 (57.4)	236 (21.3)	224 (20.8)	1,079 (100.0)

Source Korea Auto Industries Coop. Association (KAICA). 1998. Automobile Industry Handbook and author's own creation

long-term and captive business relationship with Hyundai Motor in order to become a member of the Hyundai Cooperative and receive support from the company. Table 5.2 shows that out of the total 1079 first-tier parts companies registered with the Korea Automobile Manufacturers Association in 1997, 619 (57.4%) supplied only one finished carmaker and 236 (21.3%) supplied two, of which half supplied two firms in the same *chaebol* group.

Paternalistic control by the parent company arose in the context of captive relationships in which suppliers were completely dependent on a single automaker. The head of the finished car group or car company exercised arbitrary power from a patriarchal position, even over non-affiliated parts makers that were not linked to equity ownership. Parts makers had to reliably provide parts at a low price to meet the needs of finished carmakers. These captive trading relationships and both regular and irregular price cuts severely limited the independent growth of parts makers.

5.3 Systematization of Extended Quasi-Vertical Integration in the 2000s

After growing rapidly through extended quasi-vertical integration, the Korean automobile industry experienced rapid industrial restructuring as Korea went through the Asian financial crisis in the late 1990s. In the 2000s, Hyundai Motor grew into one of the world's top five automakers, returning to rapid growth by exploiting its modular parts supply chain, improving quality management, and expanding overseas production. During this process, the quasi-vertical integration spanning the group's internal and external networks became more systematic.

5.3.1 Modular Production and Fostering Affiliate Modular Makers

As shown in Chapter 2, after establishing a monopoly system in Korea by acquiring Kia Motors in 1999, the Hyundai Motor Group began to formalize the extended quasi-vertical integration by pursuing an aggressive management strategy. In this process, the modular production system has played an important role.

Modular production enables automakers to achieve cost reductions by outsourcing a significant part of the production process, while at the same time allowing them to focus limited internal resources on core functions such as planning and designing new cars, thereby increasing flexibility and efficiency (Jo, 2001). To achieve these advantages, there needs to be a competent modular producer capable of taking charge of the production of modular parts on behalf of finished car makers.

However, it was not easy to find domestic parts makers competent enough to take charge of producing modular parts in the Korean automobile industry, as domestic auto parts makers were mostly small and medium-sized companies under Hyundai Motor's extended quasi-vertical integration structure. On the other hand, while the large overseas parts makers that entered Korea after the Asian financial crisis had a relatively high level of competence, relying on such companies entailed the risk that internal management information could be leaked to the outside, or that Hyundai Motor would become subservient to overseas modular makers.

Accordingly, to develop modular production within Hyundai Motor, the Hyundai Motor Group has actively fostered internal affiliates as exclusive suppliers of major modular parts. The company also transformed the existing single-layered parts supply system into a multi-layered supply structure centered around a core internal modular parts maker called Hyundai Mobis. This is the typical Korean *chaebol* development path that exploits internal resources in a closed style instead of building open relationships using external players. Hyundai Motor established the new auto parts affiliate Hyundai Mobis in 2000, starting with the auto parts division of Hyundai Precision Industries, a subsidiary of the Hyundai Group. The goal of this move was to transform Hyundai Mobis into a core parts maker that supplied modular components to Hyundai Motor (Kim et al., 2011).

In order to promote the growth of Hyundai Mobis, the Hyundai Motor Group concentrated its orders for core modular parts in Hyundai Mobis. For chassis modules, Hyundai Mobis had a modular production line in the warehouse of after-sales parts for export in Hyundai Motor's Ulsan plant, and exclusively supplied modular parts to the Ulsan plant under a just-in-sequence system. In addition, Hyundai Mobis' acquisition of Mando's Poseung plant led to the company becoming an exclusive supplier of chassis modules to the Hyundai Motor and Kia plants.

For cockpit modules, Hyundai Mobis had a modular production line installed in the export logistics warehouse in Kia's Hwaseong plant and exclusively supplied these modules to Kia's Hwaseong and Sohari plants. In addition, Hyundai Mobis began supplying cockpit modules to Hyundai Motor's Ulsan plant in 2003, and has since supplied these modules for most new models at the Ulsan plant. At Hyundai Motor's Asan plant Hyundai Mobis delivered modules that were then sub-assembled with instrument panels supplied by Deokyang Industries, leading to a monopoly supply structure for Hyundai Mobis.

Finally, Hyundai Mobis began to supply the front-end modules for four Hyundai Motor and Kia models from 2003, despite the existence of other parts suppliers that previously supplied such modules. As a result, Hyundai Mobis achieved the position of a semi-exclusive modular supplier in the front-end modules sector (Hyundai Mobis, 2007).[1]

Along with the intensive ordering of the module parts business, Hyundai Motor Group also concentrated its after-sales parts business in Hyundai Mobis. As a result, Hyundai Mobis took over the after-sales parts business from Hyundai Motor on February 1, 2000, and from Kia ten months later on December 1 (Hyundai Mobis, 2007). The after-sales parts business served as an important financial foundation for the growth of Hyundai Mobis during the transition period when the modular parts business was less profitable. As shown in Table 5.3, the after-sales parts business accounts for a lower proportion of sales than the modular parts

[1] In reality, Hyundai Motor placed orders for the development and delivery of individual modular parts with multiple modular components suppliers. This appears to be motivated by a combination of technological improvement and reducing the risk associated with entrusting everything to one supplier, as well as a policy consideration at the regional level (Interviews with Hyundai Mobis executives in 2008 and 2010). However, there is a significant difference in the degree of information sharing between Hyundai Mobis and other non-affiliated modular parts suppliers (Interviews with Hyundai Mobis executives in 2010). In addition, there have been measures such as ordering modular parts with low added value from non-affiliated modular component suppliers or reducing the rate of orders from non-affiliated companies (Interviews with managers at Deokyang Industries, a non-affiliated modular parts supplier in 2008).

business, but is still highly profitable. As of 2009, Hyundai Mobis' after-sales parts business sales are less than half of those of the modular parts business, but profitability (in terms of operating profit margin) is 3.3 times higher than the modular parts business. Hyundai Mobis was able to achieve stable profits by taking full responsibility for the after-sales parts business of Hyundai Motors and Kia, capitalizing upon this opportunity to establish a foundation for growth as a modular parts specialist.

At the same time, Hyundai Motor supported Hyundai Mobis to develop proficient R&D capability in a short period of time through research staff support on a group level. The origins of Hyundai Mobis' R&D capabilities are design and R&D capabilities and staff built in the process of developing finished vehicles such as the Galloper and Santa Fe in the days of Hyundai Precision Industry. However, these staff were transferred or sent to Hyundai Motor in the process of transferring the vehicle business in 1999, and some returned to Hyundai Mobis to plan and promote the modular parts business by launching the 'Auto Parts' business promotion team. In October 1999, six researchers from Namyang Research and Development Center of Hyundai-Kia Motors were transferred to Hyundai Mobis, and a full-fledged R&D system began to take shape. In April 2002, ten chassis modular parts researchers were transferred from Hyundai Motor, leading to independent design and R&D in this field. In addition, Hyundai Mobis established the Cartronics R&D Center in May 2001 to promote the research and development of core parts related to electronic information. To this end, sixty or so researchers were transferred to Hyundai Mobis from the group affiliate Bontec (formerly Kia Electronics, merged into Hyundai AutoNet in 2006) to secure initial research personnel (Hyundai Mobis, 2007).[2] The Hyundai Mobis R&D Center has also maintained daily exchanges and cooperation with the Namyang R&D Center.

[2] The Cartronics R&D Center was later merged with its affiliate Hyundai AutoNet in February 2006 to integrate the group-level electronic devices and parts business. In June 2009, Hyundai Mobis again merged with Hyundai AutoNet and the Cartronics R&D Center was incorporated into Hyundai Mobis' research system.

Table 5.3 Sales and profits by business division at Hyundai Mobis (Unit: KRW billion, %)

		2001	*2002*	*2003*	*2004*	*2005*	*2006*	*2007*	*2008*	*2009*
Sales	Modular parts and Other parts	664	1,960	3,007	4,030	5,123	5,490	5,649	6,174	7,194
	After-sales parts	1,755	1,957	2,133	2,413	2,433	2,688	2,842	3,199	3,439
Operating Profit	Modular parts and Other parts	23	56	173	266	351	299	252	410	552
	After-sales parts	343	378	423	485	435	518	572	777	870
Operating profit margin	Modular parts and Other parts	3.45	2.86	5.76	6.61	6.85	5.44	4.47	6.63	7.67
	After-sales parts	19.56	19.34	19.82	20.12	17.88	19.27	20.13	24.29	25.31

Source Hyundai Mobis Annual Report, 2001–2009 and author's own creation

5.3.2 *Modular Production and Reorganization of the Parts Supply Structure*

Since initiating full-scale modular production, Hyundai Motor has set Hyundai Mobis' status as so-called '0.5-tier supplier' that is permitted to manage sub-parts suppliers of Hyundai Motor and Kia Motors (Interviews with Hyundai Mobis executives in 2010). In other words, Hyundai Mobis was placed at the top of the parts supply hierarchy. To achieve this, Hyundai Motor strategically concentrated orders for the three core modular parts (chassis, front-end, and cockpit) in Hyundai Mobis, and Hyundai Mobis established its position as a 0.5-tier vendor by reorganizing its component supply structure around the company itself. As discussed in Chapter 1, this reconfiguration is a typical example of a *chaebol* organization monopolizing important resources through internalization. As shown in Table 5.4, Hyundai Mobis acquired or merged with existing modular parts suppliers and core parts specialists that held competitive technology, which forced these firms to either withdraw from the modular parts business and become sub-partners of Hyundai Mobis, or exit the market entirely. After taking control of these parts makers to monopolize their core capabilities, Hyundai Mobis relegated the parts makers to the status of sub-partners that provide sub-parts to Hyundai Mobis.

This process of reorganizing the parts supply structure with Hyundai Mobis located at the apex may correspond to Parkin (1979)'s 'social closure' process. By internalizing the modular parts business and blocking outside access, Hyundai Mobis exclusively shares power over the value chain and the ensuing revenue with Hyundai Motor. Non-affiliated parts makers are not allowed to achieve industrial upgrades along the value chain through entry into the modular components business. With Hyundai Mobis in control of core modular parts and components and non-affiliated parts makers reorganized into sub-partners of Hyundai Mobis, the non-affiliated parts makers are limited to developing non-core components that constitute modular parts with 'approved drawings,' or simply processing them according to 'rental drawings' (Chung, 2007). As the role of non-affiliated parts makers is limited to non-core areas, their bargaining power inevitably declines.

As the modular parts supply network began to take off and Hyundai Mobis virtually dominated Hyundai Motor's supply of major modular parts, the overall supply structure of parts became stratified. In addition to

Table 5.4 Hyundai Mobis' M&As in the modular parts business (2000–2007)

Plant	*Products*	*M&A*
Asan modular parts	Chassis, driving seats, front-ends	Mando Youngin plant (Aug 2002)
Ewha modular parts	Complete chassis, driving seats	Seojin Industry Ewha plant (Feb 2002)
Poseung modular parts	Chassis	Mando Poseung plant (Aug 2002)
Cheonan ABS, Airbags	ABS, airbags	Bosch Korea Cheonan plant (Apr 2002)
Cheonan IP	Instrument panels (IP)	Kasco Cheonan plant (Nov 2000)
Asan Plastics	Bumpers, carrier head lamp	Jinyoung Industry (Jun 2004)
Changwon	Brake systems, steering systems	Kasco (Jun 2007)
Hanam	Brake systems	Kasco (Jun 2007)
IHL	Head lamps, rear lamps	IHL (Apr 2004)

Source Author's own adaptation from Kim et al. (2011: 373)

Hyundai Mobis, Hyundai Motor developed other affiliates into core parts makers and offered them the majority of orders for modular parts and key components. Many of the first-tier parts manufacturers that used to supply parts directly to Hyundai Motor have been reformed into second-tier manufacturers that supply components to Hyundai Mobis and other affiliates.

As a result, a new type of vertical hierarchy has formed in the extended quasi-vertically integrated parts supply network between Hyundai Mobis, affiliated core parts suppliers, and non-affiliated suppliers of other parts. In addition, Hyundai Mobis and affiliated core parts suppliers have acted as a supply chain management company responsible for supplying parts to Hyundai Motor while also overseeing production by sub-parts makers.

Prior to the 2000s, Hyundai Motor's parts supply network of extended quasi-vertical integration mainly consisted of a single-tier parts supply structure, where Hyundai Motor exerted direct control over all parts manufacturers registered with the suppliers' council. However, as the modularization of production took off in the 2000s, a multi-layered and vertical supply chain management system emerged in which Hyundai

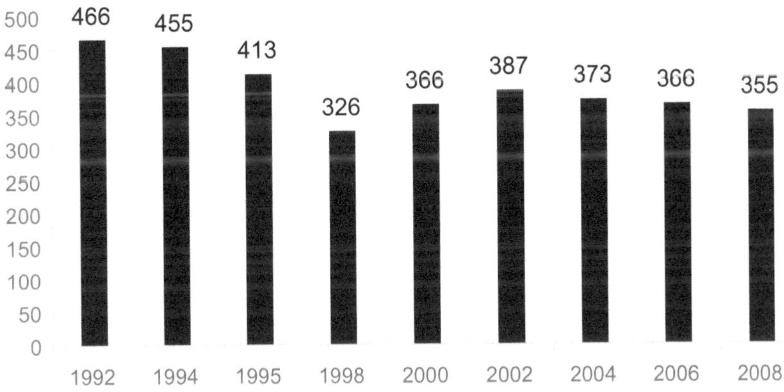

Fig. 5.1 Number of first-tier parts makers for Hyundai Motor (*Source* Korea Auto Industries Coop. Association [KAICA], *Automobile Industry Handbook*, Each Year, and author's own creation)

Motor manages affiliated key parts suppliers, who in turn oversee the sub-parts makers. The network has thus transformed into a more sophisticated entity as a result of advances in the modularization of production.

The number of first-tier parts makers for Hyundai Motor has decreased in line with the systematization of the extended quasi-vertical integration (see Fig. 5.1). From the early 1990s to the 2000s, the number of first-tier parts suppliers fell from 400 to approximately 300.[3] Moreover, the proportion of first-tier suppliers decreased in the 2000s while the percentage of second-tier and third-tier makers increased, although this was limited to small and medium-sized enterprises (see Fig. 5.2).

There is also a significant revenue gap between affiliates and non-affiliates. As Hyundai Motor Group's affiliates occupy key areas of component production, the profit gap has become entrenched due to the establishment of a multi-layered hierarchy that separates affiliates from non-affiliates (see Table 5.5).

The development of Hyundai Mobis as a core modular supplier and the reorganization of the parts supply system was born out of an attempt

[3] In Fig. 5.1, many of the 300 first-tier parts makers in the 2000s are actually '1.5'-tier parts makers that simultaneously supply modular parts makers and automakers such as Hyundai Motor. On the other hand, the sharp decline in the number of first-tier parts makers in 1998 is believed to be due to the Asian financial crisis.

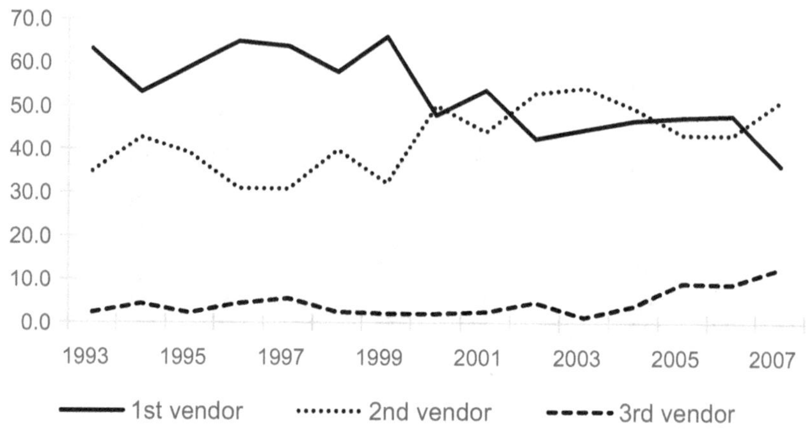

Fig. 5.2 Proportion of automobile parts SMEs in each transaction layer (*Source* Korea Federation of SMEs [KBIZ], *SMEs Survey*, Each Year, and author's own creation)

Table 5.5 Trends in operating profit margins of Hyundai Motor's parts suppliers (%)

	2001	*2003*	*2005*	*2007*	*2009*	*2011*
Hyundai Mobis	12.1	11.4	10.4	9.7	13.4	13.5
Affiliated parts suppliers (average)	6.6	8.3	5.3	3.7	7.6	5.6
Affiliated parts suppliers excluding Hyundai Mobis (average)	5.7	7.8	4.4	2.9	6.8	4.7
Non-affiliated parts suppliers (average)	4.1	−18.3	2.9	1.9	2.1	2.8

Note Hyundai Motor's suppliers are based on the list of parts manufacturers in the Korea Auto Industries Cooperative Association as of 2011. Among these companies, analysis was conducted on those for which financial data could be obtained from KIS-VALUE. A total of 231 companies are included in the analysis (9 Hyundai Motor affiliates and 222 non-affiliated parts makers)
Source Author's own adaptation from Jo and Kim (2013: 178)

to ensure Hyundai Mobis' survival rather than a plan made in advance. Hyundai Mobis was launched in 2000 with Hyundai Precision Industries as its parent. As the finished vehicle business division of Hyundai Precision Industries was integrated into Hyundai Motor, Hyundai Mobis needed to develop a new business to survive. As a result, the company put forward the proposal of supplying modular parts to Hyundai Motor, an avenue that had not been explored at the time (Interviews with Hyundai Mobis executives in 2010). This proposal aligned well with Hyundai Motor's need to build a modular supply chain. Hyundai Mobis therefore started out as a major modular parts supplier for Hyundai Motor.

Although Hyundai Mobis started out by taking advantage of an opportunity, it quickly became the center of modular parts production thanks to intensive support from the Hyundai Motor Group. With the reorganization of the parts supply structure centered around Hyundai Mobis, the stratified supply structure has continued to grow and the extended quasi-vertical integration has become more entrenched.

5.3.3 *Comparison with Toyota*

The modularization of the automobile industry, which began in Europe and later spread to the United States, is fundamentally based on the modularity of parts production (Cho, 2002). In response to the lean production system of Toyota, modular production has been promoted in order to 'use the parts manufacturers' capability to design and assemble intermediate assembly parts' as a new production method to increase production flexibility (Jo, 2001: 191). Accordingly, modular production involves the outsourcing and dispersion of production functions that were previously vertically integrated with finished carmakers. This may contribute to short-term improvement in profits through cost reductions for finished carmakers, but on the other hand it is likely to weaken the control of finished carmakers over modular parts maker due to the downsizing of internal production functions. Less control over modular parts suppliers could lead to problems in securing stable profits for finished car manufacturers. The specifics of modular production may be contingent on finding ways to overcome this trade-off between profit and control.

Toyota prefers to maintain control by minimizing modular production and refraining from setting up separate modular parts suppliers. In contrast, Hyundai Motor has been able to maintain group-level control while simultaneously leveraging modular production by using its affiliates. Toyota has been slow to adopt modularization in production, instead focusing on the modularization of design due to the relatively small wage gap between finished carmakers and parts suppliers in Japan. This means that outsourcing modular production does not lead to significant cost reductions, and finished carmakers are more efficient in terms of quality and flexibility (Ku & Oh, 2007). As a result, instead of outsourcing internal production functions to modular parts makers on a large scale through modular production, Toyota have pursued 'in-house modularization' that seeks to share parts or integrate functions within the company while maintaining existing parts supply relationships (Takeishi & Fujimoto, 2001). This method has helped to maintain the characteristics of the relationship between Japanese companies known as *keiretsu*, while not increasing the influence of external parts makers (Sturgeon et al., 2008).

In contrast, Hyundai Motor has chosen to outsource a large part of its production functions to modular parts suppliers by establishing an intermediate assembly stage known as modular parts assembly. However, instead of using external modular parts suppliers, Hyundai Motor Group has taken the approach of internalizing them by utilizing the affiliate Hyundai Mobis. Korean *chaebol* tend to deeply internalize the inertia of a modern corporate organization of 'the economies of scale and scope' in that they use resources within the *chaebol* rather than relying on external resources (Chandler, 1990). Hyundai Mobis is legally independent from Hyundai Motor, but in reality the relationship between the two is more of a closed internal transaction than an open transaction. In addition, as Hyundai Mobis is the de facto parent company of Hyundai Motor Group in terms of equity structure, it can be said that it is on equal footing with rather than subordinate to Hyundai Motor.[4] Hyundai Mobis has been able to quickly acquire high-level development and production capabilities for modular parts thanks to the support and development of Hyundai Motor Group. As a result, Hyundai Mobis exerts considerable influence on the development of new cars by exchanging information with Hyundai Motor and developing and supplying modular components that are optimized for the company's products. In this regard, Hyundai Motor's modular production differs from Toyota model in which finished carmakers Toyota lead modular production and do not foster a separate modular parts supplier.[5]

[4] As of December 2021, Hyundai Mobis was the largest shareholder, with a 21 percent stake in Hyundai Motor. In addition, the CEO of Hyundai Mobis was the head of Hyundai Motor Group and also served as the CEO of Hyundai Motor (Hyundai Mobis Annual Report, 2021).

[5] Modular production of US automakers such GM or Ford is similar to that of Hyundai Motor since it involved reducing and outsourcing production functions. However, in the United States, the reduction in production functions and outsourcing due to modular production appeared in the form of separate and independent of subsidiaries of finished automakers. Some examples are Delphi's independence from GM and Visteon's separation

5.3.4 *Quality Management and Systematic and Bureaucratic Control*

Hyundai Motor began to emphasize quality in the late 1990s as top executives were replaced in the process of changing the company's governance structure. By then, Hyundai Motor had an image that was associated with low-quality cars and was on the verge of being kicked out of the US market, which accounted for the majority of the company's exports. Under these circumstances, the new management team launched a drive to improve quality by advocating quality assurance. To this end, Hyundai Motor and Kia merged their quality headquarters in 2002 to establish a joint quality headquarters under direct management. The Quality Management Headquarters played a role in overseeing quality control across Hyundai Motor and Kia across all processes including R&D, purchasing, production, and after-sales service. The number of staff at the Quality Management Headquarters increased from 200 in 2002 to 850 in 2006 (Hyun, 2008).

A quality evaluation system was also introduced for parts suppliers under the supervision of the Quality Management Headquarters. The first system introduced was Hyundai Motor's own quality certification system called the Hyundai Quality System (HQS). This was on top of the QS9000, the criteria under which the top three US automakers select suppliers. Based on the results of HQS reviews, Hyundai Motor gave each parts maker a grade, and increased support for those selected as superior companies by shortening the payment period.

Second, the 5 STAR certification system was implemented as a quality evaluation system for first-tier parts suppliers. This involves a comprehensive evaluation of the quality, technology, and delivery of first-tier parts makers in the early stages of new car development, and gives each

from Ford. In these cases, the relationship between the modular parts supplier and the finished carmaker shifted from a strong link between subsidiaries to a weak link between independent companies. Under these conditions, it would be difficult to promote full-scale modular production akin to that of Hyundai Motor. In order to ramp up modular production, modular parts makers would have to exert a strong influence by developing and supplying modular parts optimized for the final products through the mutual exchange of information with finished vehicle makers, and this would have to take place from the development stage of new vehicles. As a result, the higher the ratio of modular production, the weaker the vehicle manufacturer's control over the modular parts manufacturer. Accordingly, unlike in Hyundai Motor, the US automakers limit modular production to a certain level and restrict the influence of modular parts companies (Sturgeon et al., 2008).

company a star rating. Companies that receive a five-star certification receive payment in cash as soon as parts are delivered, and also receive priority invitations to various events organized by Hyundai Motor.

Finally, the Supplier Quality Mark system (SQM) has been introduced for second-tier parts suppliers. This system selects specialized sub-sectors that have a major impact on parts quality and conducts on-site quality controls and inspections.

The quality evaluation system for parts manufacturers suggests that standards have become an important mechanism for coordinating and controlling the parts supply chain. The setting of standards in and of itself has a significant impact on value chain governance. According to works on standards and GVC governance structure, the lead firms of value chain define the rules and conditions of participation by setting standards or norms. This provides a framework for systematically including and/or excluding actors from participating in the value chain (Gibbon et al., 2008). The establishment of participation rules for actors becomes a key operating mechanism in value chain governance (Ponte & Gibbon, 2005). By setting standards for parts suppliers through the quality evaluation system, codifying such standards, and establishing certification procedures, Hyundai Motor has obtained greater control over parts manufacturers and is able to treat them differentially based on this system.

In fact, Hyundai Motor has divided its parts makers into the three classes of 'champion,' 'partner,' and 'candidate' based on their technological capabilities. 'Champion' parts makers are companies that have the research and design capabilities to independently perform detailed part design in line with basic specifications provided by Hyundai Motor. These are key parts makers managed by Hyundai Motor. 'Partner' parts makers have a captive relationship with Hyundai Motor and supply dedicated parts. Many of the SMEs in Hyundai Motor's extended quasi-vertical integration network fall into this class. Finally, 'candidate' parts suppliers are mainly general-purpose parts suppliers that often have a low level of technology and are excluded from support and protection within Hyundai Motor's supply network.

Prior to the 2000s, Hyundai Motor had arbitrarily exercised power over parts makers from a patriarchal position. This practice still remained in the 2000s, but at the same time, a management structure based on quality and standards was established under which hierarchical and differential management of parts makers took place through evaluations. This can be viewed as a form of systematic and bureaucratic control because it is based on rules and procedures set by the carmaker.

5.4 Factors in the Formation of the Extended Quasi-Vertical Integration

How were parts makers able to respond to Hyundai Motor's extended quasi-vertical integration network? Furthermore, how did non-affiliated parts makers that are not linked by equity ownership become absolutely committed to Hyundai Motor? How was it possible to build this network across non-affiliates?

First, the formation of oligopolies and/or monopolies in the Korean market by a small number of automakers can be cited as one explanation. When it comes to the development of an automobile industry, the more competitive the automobile market, the more equal the transaction relationships between automaker and parts makers tend to be. The inverse is also true, where more monopolistic automobile markets are more likely to lead to the development of captive relationships (Hong, 1997). Unlike in Japan, where a competitive market was formed among ten or so carmakers, Korea has maintained an oligopoly system in which the three automakers Hyundai, Kia, and Daewoo have dominated the domestic market since the early days. The oligopolistic structure of the finished automobile market implies a monopsony in the intermediate parts market. With parts demand monopolized by the three automakers, parts suppliers had no choice but to rely entirely on a small number of finished car makers.

Since the turn of the century, the monopolistic market structure of the Korean automobile market has intensified. As mentioned above, when Hyundai Motor acquired Kia in the late 1990s, the domestic automobile market transformed from an oligopoly among Hyundai, Kia, and Daewoo into a monopoly in which the Hyundai Motor Group dominated 70–80% of the total market. Accordingly, it was impossible for parts makers to seek independent survival outside of the Hyundai Motor Group. Even though they were not linked to Hyundai Motor through an equity relationship, non-affiliated parts makers had no choice but to rely on Hyundai Motor. There was no alternative but to adapt to Hyundai Mobis' monopoly on the production of modular parts in the 2000s, the reorganization of the parts supply structure, and the hierarchical and differential control stemming from the quality evaluation system. This situation created the conditions for Hyundai Motor to set limits on the opportunistic behavior of parts makers in inter-firm business relationships.

Second, the nature of Korea's path of economic development through assembly and industrialization, and the resulting low technological level of parts makers, also became an important factor that allowed extended quasi-vertical integration to become established. The industrialization of Korea, which began in the 1960s, developed from final assembly processes that are downstream in the value chain, as opposed to upstream processes such as materials and parts (Hattori, 2005). In Korea, automakers were initially created with powerful support from the developmental state, while parts companies were formed later. Automakers mainly imported core materials and parts from overseas until the 1970s, and only developed limited domestic links at the level of procuring residual parts from small and medium-sized enterprises. However, after the economic crisis in the late 1970s, there was an urgent need to procure parts at a low cost and keep them domestically sourced in a stable manner, without any reliance on foreign parts makers. By incorporating non-affiliated independent parts suppliers with a low technical capacity into the subcontracting system through the council of suppliers, automakers such as Hyundai Motor were able to build an extended quasi-vertical integration in which non-affiliated parts makers were supported through technological and managerial guidance.

For parts makers, becoming part of the extended quasi-vertical integration meant securing support and protection. Becoming a captive parts supplier for an automaker at least guaranteed a minimum number of orders on which companies could survive, even though this meant they had to supply parts at a low price. They were also able to benefit from technology, management guidance, and funding from the automakers.[6]

On the other hand, by directly fostering their own parts makers, automakers were able to overcome the huge cost burden and sourcing instability caused by importing foreign parts. From these parts makers, Hyundai Motor was able to reliably procure essential parts at a low

[6] Hyundai Motor still maintains the unique practice of 'directed sourcing' to support parts makers. This means that Hyundai Motor purchases raw materials that are difficult to procure in batches on behalf of small and medium-sized parts makers, and the SMEs process and deliver them back to Hyundai Motor. This practice has been one factor that intensifies the dependence of parts makers on Hyundai Motor and maintains the captive relationships.

price, thus saving on transaction costs. However, parts makers have been exploited through this process of reducing costs and increasing flexibility for Hyundai Motor.

The mutual interest between Hyundai Motor and non-affiliated parts suppliers allowed for the extended quasi-vertical integration to form even though it was not linked to equity ownership. This alignment of mutual interests is possible as long as automobile makers continue to grow and take responsibility for the survival of parts suppliers, ensuring the stable and timely provision of reliable parts from a patriarchal position. As a result, the extended quasi-vertical integration is built on the assumption of continued expansion and growth in production and sales.

Hyundai Motor has in fact continued to grow rapidly since the 1980s by establishing a mass production system and starting to export. This can be seen in the rapid sales growth from the mid-1980s until the 1997 Asian financial crisis. Following the crisis, Hyundai Motor achieved high growth once again through quality assurance and exploitation of the modular parts supply chain, which served as the economic foundation for maintaining and systematizing the quasi-vertical integration in the 2000s. Hyundai Motor's overseas production even led to the co-entry of non-affiliated parts suppliers included in the network called 'follow sourcing' (see Chapter 6). As Hyundai Motor continues to expand its overseas production, the production and exports of parts makers have grown in tandem. Hyundai Motor's rapid growth and expansion into overseas production allowed non-affiliated parts makers to expand their own businesses and sales, which also helped offset the pressure for flexibility and costs from Hyundai Motor.

5.5 Japan's *Keiretsu* Network and Hyundai Motor's Extended Vertical Quasi-Integration

Subcontracting systems in East Asia have received attention as network-type transaction relationships that cannot be explained by the market or hierarchy. Similar to Korea's subcontracting relationships, firms in the supply chain in Japan form a closely interdependent network that is known as a vertical *keiretsu* and includes firms ranging from automakers to parts makers. How does the relationship between automaker and parts suppliers in Hyundai Motor's extended quasi-vertical integration differ from that between Japanese companies?

Japanese *keiretsu* is characterized by a highly structured formal relationship between companies (Dicken, 2015). Inter-firm relationships are based on mutual obligations and are long-term and stable, and companies are connected through cross-shareholdings and personnel exchanges (Miyamoto, 2004).[7]

Hyundai Motor's extended quasi-vertical integration can be distinguished from *keiretsu* in that the former encompasses not only group affiliates linked by complex equity ownership relationships but also independent companies that are not. Even for non-affiliated independent companies that are free from equity relationships, Hyundai Motor exercises hierarchical control as a finished carmaker at the same level it exercises over affiliates connected through equity ownership. Furthermore, Hyundai Motor provides technical and management support in a similar manner to its affiliates. The extended vertical quasi-integration of Hyundai Motor has organized the entire supply chain in a way that hierarchically and vertically encompasses the business group's internal and external networks.

Since the introduction of modular production to the Japanese automobile industry in the late 1990s, the *keiretsu* relationships have gradually weakened. As automakers started receiving more parts as standardized modular parts to reduce costs, they became more likely to end their mutual shareholding relationships with affiliated parts makers and promote free competition among independent parts makers (Lincoln & Shimotani, 2009).[8]

In contrast, when Hyundai Motor began to ramp up module production, the extended quasi-vertical integration structure became stronger rather than weaker. By placing the group's affiliates at key points in module production and other external parts makers as their subordinates, Hyundai Motor made full use of the advantages of extended quasi-vertical integration in modular production. This contrasts with *keiretsu*, which

[7] The Japanese *keiretsu* can be divided into a horizontal form with a banking-mediated business group, and a vertical form consisting of closely linked parts supply relationships in the manufacturing industry.

[8] The Nissan Group is a typical case that exemplifies the breakdown of *keiretsu*. On the other hand, while the Toyota Group maintains more of a *keiretsu* relationship than Nissan, the nature of this relationship as a closely interdependent network has gradually weakened.

saw vertical integration weaken in response to the shift toward modular production in the auto industry.

5.6 Conclusion

This chapter has attempted to explain the inter-firm relationships that enabled Hyundai Motor's rapid growth. This model can be characterized as a parts supply system based on the concept of an 'extended quasi-vertical integration.' Hyundai Motor initially established this network by bringing not only affiliated but also non-affiliated parts suppliers into its subcontracting system in the 1980s thanks to government support. This was later systematized through Hyundai Mobis' advances in the production of modular parts and bureaucratic quality assurance in the 2000s.

As a finished car maker, the rise and fall of Hyundai Motor has the potential to determine the prospects of not only affiliated but also non-affiliated parts makers. Given this situation, Hyundai Motor was able to mobilize tremendous commitment from parts makers, even placing unreasonable demands on them. The absolute exploitation of parts suppliers' capabilities served as an important driving force for Hyundai Motor to quickly catch up to more technologically advanced companies as it began to exploit modularization and expand production overseas.[9]

External parts makers were forced to make an unlimited commitment to meet demand while absorbing the pressures and burdens placed on them by Hyundai Motor, eliminating any possibility of independent growth. However, as Hyundai Motor continued its rapid growth, parts makers were able to expand their businesses and sales, thereby offsetting the pressure of cost reductions.

As can be seen from this, continued growth in the sales of Hyundai Motor is necessary in order for the company's extended quasi-vertical integration model to work smoothly. Conversely, this implies that if the company struggles to maintain sustainable growth and scale expansion, the extended quasi-vertical integration model may be in trouble. Will Hyundai Motor be able to continue its external expansion in the future?

[9] Most people from Hyundai Motor, affiliates and parts makers interviewed for this study acknowledged that the "tremendous commitment of parts makers made a huge contribution to Hyundai Motor's rapid growth and development into a global player." (Interviews with Hyundai Motor executives in 2015).

Can the extended quasi-vertical integration model work effectively even in the face of changes in the automobile industry paradigm such as the development of electric vehicles and autonomous driving technology?

The closed nature of the extended quasi-vertically integrated supply network could also lead to limitations. The automotive industry needs to innovate, and this requires an open structure in which competent companies can collaborate in their respective areas. However, it is not easy to establish such a flexible and open arena for collaboration in Hyundai Motor's closed component supply system as it restricts outside entry.

In the face of this paradigm shift, Hyundai Motor has been called on to change. Extended quasi-vertical integration come with clear advantages and weaknesses, and it remains to be seen how Hyundai Motor will change its distinct relationship with parts suppliers in the future.

References

Chandler, A. D. Jr. (1990). *Scale and Scope: The Dynamics of Industrial Capitalism*. Harvard University Press.

Cho, C. (2002). *Advances in the Network Economy and Changes in the Parts Procurement System: Focusing on the Auto Parts Procurement System*. Korea Institute for Industrial Economics & Trade. (In Korean).

Chung, J. S. (1994). *Strategies for Nurturing Parts Makers in the Korean Automotive Industry and Responses of Parts Makers: Focusing on the Case of H Company*. Korea Development Institute. (In Korean).

Chung, M.-K. (2007). Modularization in the Auto Industry: Interlinked Multiple Hierarchies of Supplier System in Hyundai Asan Plant. *Korea Journal of Economics, 25*(3), 35–54. (In Korean).

Dicken, P. (2015). *Global Shift: Mapping the Changing Contours of the World Economy* (7th ed.). The Guilford Press.

Gibbon, P., Bair, J., & Ponte, S. (2008). Governing Global Value Chains: An Introduction. *Economy and Society, 37*(3), 315–338.

Hattori, T. (2005). *Economic Sociology of Development: Korea's Economic Development and Social Changes*. Bunshindo Publishing. (In Japanese).

Hong, J. P. (1997). Characteristics and problems of subcontractor relations in the automotive industry: Focusing on the wage gap between contractors and subcontractors. In J. P. Hong, S.-J. Cho, Y. Kim, & Y. S. Park (Eds.), *Subcontractor Relations and the Worker Gap in the Automotive Industry* (pp. 11–60). Korea Labor & Society Institute. (In Korean).

Hyun, Y.-S. (2008). The Quality Triumph of Hyundai Motor. *Journal of the Korean Production and Operations Management Society, 19*(1), 125–151. (In Korean).

Hyundai Mobis. (2007). *30 Years of Hyundai Mobis*. Hyundai Mobis Company. (In Korean).

Hyundai Motor Company. (1987). *20 Years of Hyundai Motor Company: 1967–1987*. Hyundai Motor Company. (In Korean).

Jo, H. J. (2001). Changes in the Parts Supply System according to Modularization: Focusing on the Synchronization of Production. *Economy and Society, 50*, 186–213. (In Korean).

Jo, H. J., & Kim, C. (2013). Conversion from the Captive Type to the Modular Type in the Supplier Relations: Focusing on the Case of Hyundai Motor Company. *Korean Journal of Sociology, 47*(2), 149–184. (In Korean).

Kim, C., Jo, H. J., & Jeong, J. H. (2011). Modular Production and Hyundai Production System: The Case of Hyundai Mobis. *Economy and Society, 92*, 351–385. (In Korean).

Ku, S., & Oh, J. (2007, October 26). *International Comparison of Japan and South Korea from the Viewpoint of Supplier Competitiveness: Focusing on Modularization Strategy*. Asia-Pacific Automotive Forum. (In Japanese).

Lincoln, J. R., & Shimotani, M. (2009, September 24). *Whither the Keiretsu, Japan's Business Networks? How Were They Structured? What Did They Do? Why Are They Gone?* (UC Berkeley Working Paper Series).

MacDuffie, J. P. (2013). Modularity-as-Property, Modularization-as-Process, and Modularity-as-Frame: Lessons from Product Architecture Initiatives in the Global Automotive Industry. *Global Strategy Journal, 3*, 8–40.

Miyamoto, M. (2004). *A Theory of Corporate System*. Shinsei-sha. (In Japanese)

Parkin, F. (1979). *Marxism and Class Theory: A Bourgeois Critique*. Tavistock.

Ponte, S., & Gibbon, P. (2005). Quality Standards, Conventions and the Governance of Global Value Chains. *Economy and Society, 34*(1), 1–31.

Sturgeon, T. J., van Biesebroeck, J., & Gereffi, G. (2008). Value Chains, Networks and Clusters: Reframing the Global Automotive Industry. *Journal of Economic Geography, 8*, 297–321.

Takeishi, A., & Fujimoto, T. (2001). Modularization in the Auto Industry: Interlinked Multiple Hierarchies of Product, Production, and Supplier Systems. *International Journal of Automotive Technology and Management, 1*(4), 379–396.

CHAPTER 6

The Overseas Production Network—'Model Factories' and the Global Transfer of Hyundai Motor's Production System

6.1 Introduction

Hyundai Motor has actively expanded overseas production in the twenty-first century. In order to overcome the limitations of being a middle-ranked carmaker and achieve economies of scale, it was necessary to not only export domestically produced vehicles, but also directly produce them overseas. Ramping up overseas production was also necessary in order to overcome the restrictions of trade barriers and exchange rate fluctuations that accompany export growth. Hyundai Motor constructed its first overseas plant in Alabama in 2003, followed by plants in Europe and South America. As of 2022, Hyundai Motor Group runs overseas plants in ten countries around the world. Overseas production capacity stands at 5.86 million units, accounting for 64.1% of the total production capacity of 9.14 million units (see Table 6.1).

The construction of overseas plants involves setting up production facilities overseas and procuring parts to manufacture vehicles through local employees. To date, Hyundai Motor has successfully operated overseas plants in most regions except China.[1] This means that the company

[1] The direct cause of the decline in Hyundai Motor's market share in China was the political fiasco over the deployment of the THAAD US anti-missile defense system in Korea in 2017. However, Hyundai Motor Group's Chinese market strategy of positioning products between American and Japanese high-end cars and Chinese mid- to low-end cars

H. J. Jo et al., *Agile Against Lean*,
https://doi.org/10.1007/978-981-99-2042-6_6

Table 6.1 Production capacity of Hyundai Motor Group (as of 2022)

			Production Capacity	*(%)*
Domestic		Hyundai Motor	1,800,000	35.9
		Kia Motors	1,480,000	
		Sub-total	3,280,000	
Overseas	China	Hyundai Motor	1,250,000	64.1
		Kia Motors	890,000	
	India	Hyundai Motor	680,000	
		Kia Motors	300,000	
	Czech Republic	Hyundai Motor	350,000	
	Slovakia	Kia Motors	350,000	
	Turkey	Hyundai Motor	230,000	
	Russia	Hyundai Motor	250,000	
	USA	Hyundai Motor	370,000	
		Kia Motors	360,000	
	Mexico	Kia Motors	400,000	
	Brazil	Hyundai Motor	180,000	
	Indonesia	Hyundai Motor	250,000	
		Sub-total	5,860,000	
		Total	9,140,000	100.0

Source Hyundai Motor and Kia Motors websites and author's own creation

has succeeded in efficiently transferring the production system established in Korea to overseas locations.

In this chapter, we examine how Hyundai Motor's domestic production system has been transferred and operated overseas in line with the expansion in overseas production. The production system was originally set-up within the institutional conditions of Korean society, and has a certain path dependency. Successfully running plants in other countries requires efficiently relocating Hyundai Motor's production system while also adapting to local circumstances.

Abo (1994) explained the establishment of overseas plants through the concepts of 'applying' and 'adapting' production systems. When a production system established in the home country is transferred to an overseas location, certain parts are applied without change, while

has also failed over the mid- to long-term (Interviews with a Hyundai Motor executive in 2022). As a result, Hyundai Motor's production capacity has declined from an annual unit of 2.7 million in 2016 to an annual unit of 2 million now, and market share has fallen from 7.9% in 2016 to 1.7% in 2022.

others have to adapt to the local conditions. The details of application and adaptation and the degree to which they are mixed will naturally differ depending on the path dependence and strategy of the finished car manufacturer.[2]

In the previous chapters, we noted that Hyundai Motor's production system can be understood as an agile production system featuring flexible production technology and skill-saving work organization led by engineers. How has this production system been transferred to overseas plants with conditions that differ from Korea? And how has this system been applied and adapted to local conditions?

This chapter first examines how Hyundai Motor transferred its agile production system overseas, with a focus on the concepts of 'application' and 'adaptation.' The company has sought to standardize the agile production system set-up in Korea and apply it to green-field plants overseas.[3] However, certain adaptations have also been made in the management of local staff. We also explain the 'follow sourcing' of Korean parts makers that went hand in hand with the establishment of overseas plants. Second, we investigate how overseas plants have been run from the perspective of the global value chain. As a multinational company, it is appropriate to approach Hyundai Motor's overseas production by focusing on the headquarters-subsidiary relationship at the global rather than national level. In other words, we tackle new car development at head offices and the daily operations of overseas plants separately. This chapter also looks at the human resource management and parts supply systems of overseas plants. Finally, we compare the overseas strategies of Hyundai Motor and Japanese carmakers such as Toyota.

6.2 Transplanting the Agile Production System Around the World

Hyundai Motor's establishment of overseas plants has involved transferring the company's agile production system to overseas green fields and applying the system with as few changes as possible. At the core of this strategy is the concept of a 'model factory.' The skill-saving work

[2] Abo (1994) compared and studied how the production systems of Japanese automakers became mixed when they were transferred overseas.

[3] The green-field plant is a factory which is erected on a previously underdeveloped site, in contrast with an extension or conversion of an existing plant (Black, 1997: 203).

organization is well-suited to this purpose, although a certain degree of adaptation has been required with regard to the localization of human resource management.

6.2.1 *The 'Model Factory' Concept and Construction of Green Field Plants*

Let's begin by taking a look at the construction process for overseas plants. The Hyundai Motor Group has adhered to a single investment scheme in the construction of overseas plants. This was partly because it was difficult to find joint venture partners due to the company's relatively low profile during the early stages of entry, but also because of negative past experiences in joint ventures with advanced automakers such as Ford, Mitsubishi, and Mercedes. In the 2000s, Hyundai Motor's direct investment which expanded in full swing was an independent investment with 100% sole ownership. Hyundai Motor's overseas production was a preventive strategy to respond to uncertainties such as trade friction and exchange rate fluctuations, but can also be characterized as an aggressive investment to actively expand sales volume in overseas markets and enhance the company's brand image (Interviews with a Hyundai Motor executive in 2015). This contrasts with the situation in the mid-1980s when Japanese automakers were effectively forced to build overseas plants due to pressure from trade friction (Jo & Jeong, 2016a).

Hyundai Motor's direct investment is aggressive in nature, which needs to be understood in relation to the characteristics of corporate governance. Hyundai Motor Group controls all affiliates with strong authority through roundabout and cross-shareholding, as explained in Chapter 2. The chairman of the company, or *chongsu,* owns a 7% stake in Hyundai Mobis, the parent company.[4] The chairman of the group has led major decision-making related to overseas investment, and served as a strong driving force behind Hyundai Motor's rapid growth. Fujimoto (2006) pointed out that the organizational strength of Korean companies lies in the bold investment and modular architecture of the *chaebol* system.

[4] As discussed in Chapter 2, the chairman of the Hyundai Motor Group exercises strong leadership through the roundabout and cross-shareholding structure among group affiliates. Despite owning only 6.96% of Hyundai Mobis and 5.17% of Hyundai Motor, the chairman of the Hyundai Motor Group controls the entire group through this roundabout and cross-shareholding structure, acting as the *de facto* owner of the entire business group.

In contrast, overseas production by Japanese automakers has been more gradual. For example, Toyota went through a step-by-step development process from exports to joint ventures and sole proprietorships in Europe. Toyota has maintained a prudent investment strategy under a system of professional management, which shies away from aggressive investment decisions. The countries Toyota has invested in include Denmark, the United Kingdom (single venture, 1992), France (single venture, 2001), Poland (joint venture, 2005), and the Czech Republic (joint venture, 2005) (Guo & Upathamwaranon, 2008).

If we focus on the concept of production systems, the goal of building overseas plants was to transfer the agile production system established in Korea overseas. What is noteworthy here is the concept of a 'model factory' with an annual production capacity of 300,000 units. Since 2002, the Hyundai Motor Group has been through significant trial and error in building the first overseas plant in Alabama. This plant was modeled after the Asan plant in Korea, as discussed in Chapter 2, but experienced a number of difficulties in stabilizing production due to the high degree of automation and advanced production technology. During this process, the concept of the 'model factory' was created.

Hyundai Motor Group first established the model factory concept during the construction of the US Georgia plant in 2006, and the Czech plant constructed in 2007. A model factory is a plant with standardized production facilities and equipment and has an annual production capacity of 300,000 units. Hyundai Motor created this model to maximize investment efficiency, taking into account non-operating factors and product quality history. In other words, the company standardized its manufacturing design technology so that it could be applied across all overseas plants.

Hyundai Motor's model factory seeks to minimize adaptation problems that could occur in local plants by setting up production facilities and equipment in Korea based on a standardized factory layout. The construction of overseas plants is led by the head office's Manufacturing Design Technology Center. All molds used in the body, painting, and final assembly processes are developed and manufactured in Korea, and all equipment is also ordered domestically and tested before being transported to overseas plants, assembled, and subjected to a final inspection through facility test runs (Interviews with a Hyundai Motor executive in 2021). The model factory model has been completed at the third factory built in Beijing, which entered operation in 2014. The Manufacturing Design Center in Korea subsequently created a rule book, and all overseas factories built thereafter applied the model factory model.[5]

When Hyundai Motor promoted overseas production, the most pressing question was how to effectively use the manufacturing design technology established in Korea in overseas plants. The model factory system is most appropriate when the local conditions abroad are ripe for green-field investment. Since there is almost no infrastructure such as local regulations or unions in such locations, it is possible to transfer the production system from the home country with minimal adaptation.

Green fields are attractive locations for Hyundai Motor to invest in. The company chose these regions for building overseas plants so that the production system could be efficiently transferred with a minimal level of influence from the local environment. This contrasts with brownfield investment, where it is easier to secure skilled labor. Most of Hyundai Motor's overseas plants are located in green field rather than brownfield areas. The plants in Alabama, Georgia, and Slovakia were all built on farmland a certain distance from large cities and assisted through unconventional support from local governments such as free land and long-term corporate tax exemptions.

6.2.2 *The Application and Adaptation of Human Resource Management*

In addition to standardized manufacturing design technology, the skill-saving work organization also played a large role in the successful transfer of Hyundai Motor's production system. Since the agile production system is less dependent on skilled labor, overseas plants have not had much trouble hiring workers in green-field areas. Due to the high degree of

[5] The concept of the model factory has weakened somewhat in recently constructed small-scale overseas plants, such as the Brazil plant (annual production of 180,000 units) and Indonesian plant (annual production of 250,000 units). Compared to the period of rapid expansion in overseas production before the mid-2010s, Hyundai Motor's growth has slowed down and appears to go through a certain period of adjustment. As the global automobile market becomes more consumer-centered, production flexibility is now more important than economies of scale through model factories. Equipment and tools standards remain important for overseas plants, but have been accompanied by certain variations in the automation rate, tools, and equipment makers in recent times (Interviews with a Hyundai Motor executive in 2022).

Table 6.2 Employees at Kia Motors Slovakia in 2015

Job	*No. of Employees*	*Position*	*No. of Employees*	*%*
Managers	442	Manager or higher	51	12.6
		Assistant Manager	122	
		Specialist	269	
Workers	3,026	Workers hired from employment agencies not included	274	86.2
Expatriates	43			1.2
Total	3,511			100.0

Source Interviews with a Kia Motors Slovakia's executive in 2017 and author's own creation

automation, a skilled workforce with a lot of experience was not required. In this way, the model factory concept has a 'selective affinity' for skill-saving labor because the success of the system does not hinge on active participation from workers.[6]

What is the composition of the workforce at Hyundai Motor's overseas plants? With the exception of a small number of corporate heads, executives, and expatriates, the vast majority of workers are locally hired. For example, the percentage of locals at Kia Motors' Slovakia plant was 98.8% in 2015, with expats dispatched from Korea accounting for only 43 of the total 3511 employees. Management positions account for 12.6% of the total while production workers make up 86.2%, with the latter accounting for the majority of the employees (see Table 6.2).

The performance of overseas plants depends on efficiently managing local staff by helping them adapt to Hyundai Motor's organizational culture. As a result, the characteristics of Hyundai Motor's organizational culture are also reflected in human resource management at overseas plants. Blue-collar workers, who make up the majority of the local workforce, are mostly unskilled people hired from nearby agricultural or service industries. Rather than seeking out high-skilled workers, Hyundai Motor aims to hire workers who have a low level of skill but are hardworking and able to avoid repeated mistakes (Jo & You, 2011). This is why the skill-saving work organization is suited to the company's standardized manufacturing design technology.

[6] Hyundai Motor's production system evolved under Korea's institutional conditions and labor–management relations, and has considerable path dependence.

However, the managerial positions that account for around 10% of overseas plant employees do require specialized skills that differ from production workers. Production management, HR and labor management, and financial accounting skills are required to be in charge of the daily operations of overseas plants. This means the local conditions and culture have to be taken into account in order to employ and utilize high-quality local staff in managerial positions. In other words, a certain degree of localization or 'adaptation' is required.

As there was no clear HR policy in place when Hyundai Motor first ventured overseas, it was inevitable that there would be a certain amount of trial and error in hiring local managers. At the time, the company thought it would be possible to apply the HR system from Hyundai Motor's headquarters with regard to salary and promotions, but soon realized this was not so easy. Understanding the local situation required researching labor laws and market conditions through lawyers. Hyundai Motor learned that the hierarchical level system and promotions based on seniority from the head office could not be applied to overseas plants without some modifications. During this process, the local HR manager was replaced several times and there was considerable trial and error in revising company regulations to suit reality.

This was the background to Hyundai Motor Group creating the 'Global Human Resource Standard' (GHRS) in 2006 with the help of a consulting firm. However, the GHRS is not a strict set of rules that is applied equally to all overseas subsidiaries including overseas plants. Although there are common norms for human resource management in overseas plants, the rules are applied in a loose fashion that takes into account the local conditions where each plant is located.[7]

In summary, as we have seen so far, Hyundai Motor Group's direct investment and construction of overseas plants involved aggressive investment by top management and creating the model factory concept to target overseas markets. The company transferred its agile production system to overseas green-field locations, and recruited low-skilled production workers required to run the plant. However, in order to hire

[7] Even after being revised in 2011 to conform with the revision to Hyundai Motor Group's core values, the GHRS is still treated as a set of flexible guidelines for integrating human resource management at overseas subsidiaries.

high-quality local managers, the group created the GHRS and adapted HR management policies to a certain extent.

6.2.3 'Follow Sourcing' by Korean Parts Suppliers

One notable point is that the construction of Hyundai Motor's overseas plants has often been accompanied by Korean parts makers entering the local market. This practice of 'follow sourcing,' in which Hyundai Motor uses the same suppliers from Korea in different locations or plants, was another key factor in the early success of the company's overseas plants. In the early 2000s, the Hyundai Motor Group assumed it would not be possible to receive deliveries from the local parts makers in the host country under favorable conditions due to the company's lack of recognition and small production capacity of 2 to 3 million units per year. In this context, Hyundai Motor had no choice but to ask Korean parts makers who were already familiar with the company's business practices to enter the market together.

Due to the preexisting culture of vertical business relations, Korean parts manufacturers had no choice but to closely cooperate with Hyundai Motor in the process of design changes arising from errors that occurred in the process of new car development. Hyundai Motor attempted to lower transaction costs by maintaining the practices of transaction relationships set-up in Korea overseas (Interviews with a Hyundai Motor executive in 2015). Accordingly, the company encouraged not only business group affiliates but also major non-affiliated first-tier components manufacturers to enter the market together, a request with which most of them complied.[8]

The successive construction of overseas plants appears to be closely related to the production volume of Korean parts makers that entered the market together. There was a need to secure a large-scale supply of parts for multiple finished car plants in order to guarantee profits for Korean

[8] Beginning with the installation of production lines by parts makers in the early stages of the Korean automobile industry, Hyundai Motor has helped parts makers grow by providing guidance on quality and productivity improvements and even providing education on cost calculations and bookkeeping legislation, as shown in Chapter 5. As a result, it was unthinkable that parts makers would refuse Hyundai Motor's request for follow sourcing. Even in the early stages of the overseas venture, Hyundai Motor passed on expertise in local licensing, taxes and administrative processing (Interviews with a Hyundai Motor executive in 2015).

parts makers that entered the market for the sole purpose of supplying Hyundai Motor Group. For example, the Czech plant and Slovakia plant are relatively close to each other in Europe, and the US Alabama plant and Georgia plant are also not too far apart. This made it possible to achieve economies of scale by allowing parts suppliers located in the center to supply both plants.

For example, there are a total of 33 plants owned by Korean national parts makers that entered Europe in the mid-2010s, including first and second-tier parts makers. Nineteen of these firms entered when Kia Motors built a local plant in Slovakia, while the entry of the rest coincided with the launch of Hyundai Motor's Czech plant. The follow sourcing firms include the affiliates Hyundai Mobis, Hyundai Dymos (currently Hyundai Transys), and Hyundai Hysco (currently Hyundai Steel), as well as non-affiliated parts makers that deliver to the Hyundai Motor Group in Korea, such as Halla Climate Control (currently Hanon Systems), Sejong Industrial, Hanil E-hwa (currently Seoyon E-hwa), Donghee, and Sungwoo (Interviews with a Hyundai Motor executive in 2015). Looking at the transaction-by-transaction structure of the 25 plants that responded to our survey, there were 12 first-tier parts makers, six that fall into both the first-tier and second-tier category, and seven in the second tier (see Table 6.3). Second-tier parts makers may also be seen as belonging to the category of follow sourcing in that they had entered or predicted to enter Europe, following Hyundai Motor's entry into the European market.

As mentioned above, in the case of Europe, most of the plants belonging to Korean parts makers engaged in follow sourcing are

Table 6.3 Korean automotive parts suppliers responsible for follow sourcing in Central and Eastern Europe

	No. of firms	*Country*		
		Czech Republic	*Slovakia*	*Poland*
First-tier suppliers	12	4	5	3
Both first-tier and Second-tier suppliers	6	5	1	0
Second-tier suppliers	7	2	2	3
Total	25	11	8	6

Source Survey results from Korean auto parts suppliers responsible for follow sourcing in Europe and author's own creation geographically located between two of Hyundai Motor Group's finished car plants, with many in the Czech Republic (11), Slovakia (8), and Poland (6). These firms have maximized economies of scale by engaging in cross-transactions with Hyundai Motor Group's two finished car plants in Slovakia and the Czech Republic (Interviews with a Hyundai Motor executive in 2015).

Hyundai Motor's practice of follow sourcing for overseas plants is different from that of Japanese automakers, which receive most of their parts from local parts makers in Central Europe. At Toyota Peugeot Citroen Automobile Czech (TPCA), a 50:50 joint venture between Toyota and Peugeot, Toyota and Peugeot affiliates each account for only 34% of parts suppliers, with the remaining 32% comprised of European parts makers. Toyota prefers deal with European parts makers in order to absorb advanced technologies in the early stages market. Japanese parts makers tend to favor joint ventures with local parts makers because it allows them to mitigate the risk of investing alone (Jo & Jeong, 2016a).

6.3 Operating Organizational Capabilities on a Global Scale

The operations of Hyundai Motor Group's overseas plants can be better understood from the perspective of a global value chain around the headquarters than on an individual plant level. At the global level, the secret to success is that Hyundai Motor Group optimizes and operates organizational capability, as discussed in Chapter 1. Organizational capability refers to the ability to utilize tangible and intangible resources as a dynamic capability unique to a company (Nobeoka, 2007: 4).[9] How does Hyundai Motor Group optimize its organizational capacity at the global level? The operations of overseas plants can be broadly explained through the two dimensions of new car development and daily factory operations. The former is centered on the head office, while the latter is carried out through efficient management of local staff.

[9] In other words, the organizational capability of a company refers to technical resources, human resources, and an organizational routine that is a common behavioral pattern that integrates and utilizes them efficiently.

6.3.1 *Concentrating New Car Development at the Headquarters*

The Hyundai Motor Group has integrated 5–6 vehicle platforms and developed all new cars for Hyundai Motor plants around the world at the Namyang R&D Center. The company has made great progress in standardization, and modularization has also advanced by more than 40%, greatly improving the ease of parts development and vehicle assembly. The Namyang R&D Center is fully responsible for the development of all new cars, from globally standardized mass-produced models to regional-specific models that are localized and produced at overseas plants. The majority of the vehicles produced at overseas plants are specialized models that are based on the mass-produced models but have been modified by incorporating the demands of local customers. In other words, car bodies and designs are often developed as regionally specialized models that reflect the needs of local customers by region. Regionally specialized models refer to develop specialized models that are differentiated in design, body size, interior and parts, etc. by reflecting the needs of local customers. For example, the Czech plant, the European Design Center, and the European Technology Research Institute located in Frankfurt all support the development of products that are optimized for Europe in terms of driving performance and the durability of parts. However, their role is limited to assisting in the development of new cars at the Namyang R&D Center.

Preparatory work for the production of new models at overseas plants is also centered around the head office. The entire process of ordering, installing, and preparing production facilities is led by professional staff from the manufacturing design center at the head office. When Hyundai Motor first set-up overseas plants in 2006, 200 or so technical support staff were dispatched from the headquarters to prepare for plant construction and the mass production of new models while also improving the organizational capabilities of overseas plants. This reliance on the head office has decreased over time as overseas plants build up their own production experience. At present, when an overseas plant prepares for the mass production of a new model, the project is supported by around 50 to 60 technical staff from the head office's manufacturing design technology center. This includes 20 engineers and 30–40 production workers in charge of setting up fixtures, tools, and molds.

Let's take a closer look at this preparation process (see Fig. 6.1). When a blueprint for a new model comes out from the Namyang R&D Center,

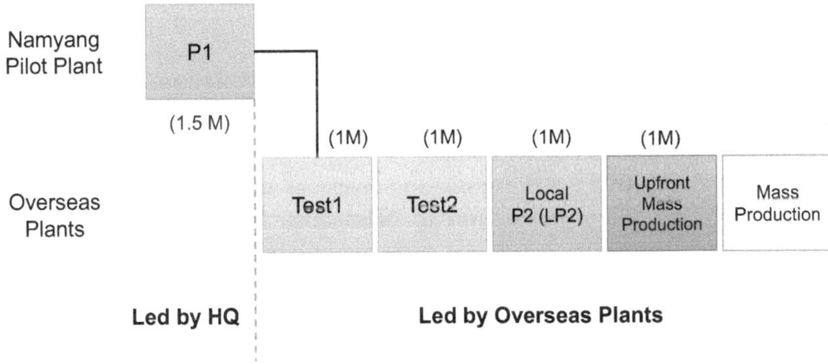

Fig. 6.1 Preparatory procedures for mass production at Hyundai Motor's overseas plants (*Note* M and HQ indicate months and headquarters, respectively; *Source* Interviews with Hyundai Motor in 2022 and author's own creation)

pilot cars are produced at the pilot center located within the R&D center to identify and solve potential problems in mass production of the new model. The pilot center is a large-scale facility that resembles a mass production plant, and serves as a cornerstone of Hyundai Motor Group's new car development process.[10]

In the pilot production stage, 40 or skilled workers from overseas plants visit the Namyang Pilot Center on three occasions for training courses that each last around two weeks. The workers check the assembly method created by the manufacturing design technology center, and if they find a problem while producing the pilot car, they reflect it in the design and make amendments. The first stage of pilot production (P1) takes place at the pilot center in Namyang, while the second stage (P2) occurs in the local overseas plant. Since changes in facilities and tools may arise during the transportation period that takes two months to the plant, it is necessary to conduct pilot production (LP2) that corresponds to the P2 stage at the local plant.[11] The local plants test-run facilities and

[10] Details of Hyundai Motor's Pilot Center have been described in Chapter 3.

[11] This is different from the new car development process at domestic plants, which connect to the plant in charge of mass production after both stages of pilot production (P1 and P2) in Namyang are completed. See Jo and Jeong (2017) for more on the new car development process at domestic plants.

tools transported from Korea and are placed in charge of pilot production (LP2). P1 takes a month and a half, while the remaining stages take a month each.

At overseas plants, the T stage is the same as at domestic plants. Facility trial operations consist of two stages, T1 and T2. This is a new process adopted in the 2000s for domestic and overseas mass production plants, and comes after pilot production in Namyang. Even if a problem has been solved at the pilot center, it is necessary to check the equipment and quality again while producing the new model under different conditions in the overseas plant. Domestic plants require a total of 18 months from model selection decision to mass production, while overseas factories take 20 months, including two months for transporting equipment and tools. What is notable here is that the design of the new model and production of pilot cars are performed in Korea, while the remainder of the process is done locally, but all processes prior to mass production are led by the head office.

As described above, new car development at Hyundai Motor's overseas plants has been carried out in a global manner by leveraging the organizational capabilities of the company's head office. Product development is mainly handled by the Namyang R&D Center, while overseas design centers and technology research centers incorporate the demands of local customers. The head office's pilot center prepares for mass production at overseas plants while solving problems through the production of pilot cars, then the manufacturing design technology center places orders for molds and equipment, completes trial runs, and then transports and installs these facilities in overseas plants. When it comes to parts, Korean parts suppliers that have entered the overseas market together either develop parts domestically and then export them, or produce them locally to supply to the overseas plants. In summary, the capabilities of the head office are utilized for non-routine aspects such as new product development, preparing for mass production, and parts development, while overseas plants demonstrate their capacity to the extent that they ensure these processes are carried out without a hitch.

6.3.2 *Daily Plant Operations*

Most of the technical support staff dispatched from the head office return home once mass production of the new model has achieved a level of stability, and overseas plants are given a high degree of autonomy through

Table 6.4 Characteristics of production processes at Kia Motors Slovakia

Process	*Main features*
Press	KUKA robot installation, small parts supplied by parts suppliers, produces 17 parts and 6800 panels
Welding	5 models (Ceed 3-door, 4-door, Wagon, Venga, Sportage) produced from 2 main bucks, up to 8 models available, moving parts, body, roof, and reinforcement welding
Painting	Automation rate of 67%, rotation dip method, PBS of 120 units, CTS of 40 units
Final Assembly	2-hour shift for skilled workers only within a work organization; skillet conveyor; 50 sequence items (70% of total) and automatic supply in sequence from stack

Source Interviews with Kia Motors Slovakia's executives in 2017 and author's own creation

the localization of daily operations. Let's take a closer look at the everyday operations of overseas plants.

Once mass production begins, the overseas plant is responsible for daily operations. Each plant is expected to reach the target pass rate of 95% within one to two weeks after beginning mass production of a new model. Once this is achieved, all technical support staff dispatched from the headquarters return to Korea, and local engineers take over daily operations. Now that Hyundai Motor has more than 15 years of experience in setting up and running local plants, the head office no longer plays such a hands-on role. Local staff take charge of everyday operations, including maintenance and quality control, and request technical support from the head office only when necessary.

For example, the plant in Slovakia currently produces Ceed, a European-specific C-segment product, Venga, a B-segment product, and Sportage, a compact SUV. Since Ceed has three models (3-door, 4-door, and wagon), a total of five vehicles are produced in a flexible manner from two main bucks.[12] The plant in Slovakia is equipped with state-of-the-art production facilities. An outline of the production process is shown in Table 6.4.

The current production speed at the Slovakia plant is '62 UPH,' with 62 vehicles produced per hour. Since 2012, the plant has operated on

[12] Mixed production in which parts for different types of vehicles are assembled on the same line is made possible by the main buck, which rotates and supplies parts for different vehicles in three dimensions.

a three-shift system. Each group works for 7.5 hours with 30-minute breaks in between, so the daily operating time is 22.5 hours. The operation rate is close to 100% as there are few disruptions to production. Since the Slovakia plant has been in operation for more than 15 years and the average length of seniority among local employees exceeds 10 years, over time there has been a growing amount of improvement activities to make production more effective (Interviews with a Kia Motor Slovakia's executive in 2018). Best practices with regard to improvements at overseas plants are immediately recorded and applied to other overseas plants to achieve the same results.

However, this does not mean that overseas plants are completely autonomous. When a problem arises that is difficult for local engineers to solve, the issue is solved through an online meeting with engineers from the manufacturing design technology team at the head office. As production facilities are monitored with cameras, a video conference is held to solve the problem. If this fails to resolve the issue, engineers from the head office will travel to the plant in person to solve the problem (Interviews with a Kia Motor Slovakia's executive, 2022).

In summary, Hyundai Motor Group's overseas plants continue to depend on the head office for the development of new models, although day-to-day operations are localized and autonomous to a large extent. In this way, the success of overseas plants has come about by combining the organizational capabilities of the head office with the localization of daily operations at an optimal ratio.

6.3.3 *Human Resource Management for Locally Hired Staff*

Overseas plants have become responsible for daily operations due to the accumulation of management experience among locally hired staff. In the early days of the overseas venture, expats dispatched from the headquarters were mainly responsible for plant operations. Today, however, locally employed managers are in charge of day-to-day operations. Although the heads of overseas plants and key offices are still Korean executives from the head office, 92% of key positions are held by locally recruited managers.

New production workers receive group training for 3–5 days after joining the company, and are then assigned to a department for on-the-job training. On average, it takes around six months for new employees to become skilled at their jobs. Each employee works eight-hour days, including one 10-minute break and a 20-minute lunch break, and the plants run on a three-shift system.

Workers in overseas plants perform job rotation within a working group, eventually making them skilled at 3–4 jobs. Evaluations are conducted every three months. However, the goal of becoming multi-skilled is not to build a high degree of skill, but to reduce the monotony of work and to cope with absences. Accordingly, Hyundai Motor Group has transferred its 'skill-saving' work organization from domestic to overseas plants. Since most problems are already solved during the product design stage and parts delivery, a high level of skill is not required for manufacturing finished vehicles. Instead, workers are simply required to avoid repeated mistakes and demonstrate a hardworking attitude (Jo & You, 2011).

One interesting point is that overseas plants have comparatively good conditions for promoting systematic skill formation among shop-floor workers due to the absence of strong unions. However, there are no overseas plants that encourage systematic skill formation outside of having workers become proficient in 3–4 different jobs. The path dependence of the skill-saving work organization that evolved in Korea has spilled over to overseas plants.

When a production worker first joins the company, they begin as a general operator. After at least three years, they are eligible for promotion to the senior operator through performance reviews, and can then be promoted to supervisor after another three years. However, there are a limited number of promotion spots. As of 2016, the annual average turnover rate was 8.4% across all overseas plants, but this differs depending on local circumstances. Labor-management relations at overseas factories are more congenial due to the absence of strong unions. Most local plants do not have unions, and strikes are rare even when a union does exist. After learning from the failures of confrontational labor–management relations at the head office in Korea, overseas plants have attempted to identify and resolve employee complaints on an individual level.

It can be seen that human resource management of production workers and managers of overseas plants has little to do with each other. This is why there is no pathway for production workers to be promoted to a managerial position. Wages for production workers are hourly, consisting of a 70% basic wage and 30% variable wage, which is a higher basic wage proportion than at the head office.

The personnel system for managerial positions in overseas plants has adopted more local practices after going through trial and error in the early stages. The GHRS only offers loose guidelines for managing locally hired staff at plants around the world, and individual overseas plants have adapted to local institutional conditions. For example, Hyundai Motor initially tried to copy the personnel system of the head office when European subsidiaries were first established. However, this attempt was soon abandoned, and a European personnel system gradually took root. With greater job mobility in Europe than Korea, the biggest challenge in human resource management has been how to secure excellent human resources by accepting the job performance-based pay system to facilitate recruitment of managerial positions and personnel management. Although the managerial rank system appears to be similar to that of the head office in Korea, in practice there are more elements of European-style HR systems, such as job performance-based pay, continuous salary level upgrading within specific jobs, and pay adjustments to market prices (Jo & Jeong, 2016b).

Nevertheless, Korean expats dispatched from the head office still play an important role in the daily operation of overseas plants. In the early days, expats held almost all key position, but the proportion and role of expatriates as 'coordinators' have now been reduced. The average number of expat employees has fallen from 65 to 40. Korean employees from the head office now assist locally recruited managers and participate in appraisals, reviews, and the approval of new local employees.[13]

In short, there is a marked difference in the HR policies for shop-floor workers and managerial positions at overseas plants. Hyundai Motor applied the skill-saving work organization from Korea to the shop-floor, but widely accepted local practices such as job-based or job-evaluated pay systems have been adopted for managers in order to attract better talent. In this way, the company has adapted to local institutional conditions.

[13] Local Korean employees from the head office are dispatched from each department, including research, quality, manufacturing design technology, and purchasing in order to maintain close communication with the headquarters. When an official letter comes from the head office, they report and respond to the head of the local plants at overseas.

6.3.4 *The Parts Supply System*

How are parts supplied to produce finished cars in overseas plants? Most core parts are developed by Korean parts suppliers that entered local markets alongside Hyundai Motor through the practice of follow sourcing. These parts are either imported by knockdown (KD) from Korea or another country, or are produced and delivered by the local plants of parts manufacturers. For example, Table 6.5 shows that the nominal localization rate of parts sourced to Kia Motors' Slovakia plant is 45.5%. However, the actual degree of localization is lower than this. According to the table, 51.1% of parts procured by all parts makers in Europe are KD and come primarily from Korea, with some from India and China. The remaining 45.5% are sourced from European parts makers. However, following sourcing by Korean parts suppliers accounts for approximately half of this figure, so the actual degree of localization is only slightly above 30% (Jo & Jeong, 2016a: 61–63). The procurement rate of components produced by purely foreign parts makers stands at around 21.5%.

As we have seen so far, localization has generally been successful in the daily operations of Hyundai Motor's overseas plants, and the degree of localization has increased considerably over time. In the day-to-day operation of producing vehicles according to production plants, proactive

Table 6.5 Local procurement rate of Korean auto parts makers responsible for follow sourcing in Europe (as a percentage of total purchase amount)

		Locally sourced		*KD imports from Korea*	*Others*
		Follow sourcing by Korean firms	*Foreign parts suppliers*		
Total		24.0	21.5	51.1	3.4
Entry period	Before 2006	21.1	26.1	46.6	6.2
	After 2007	29.3	14.7	55.0	1.0
Position in the supply chain	First-tier	22.8	23.4	51.9	1.9
	Both first and second-tier	22.2	23.0	45.7	9.2
	Second-tier	27.7	17.2	55.2	0.0

Source Author's own adaptation from Kim and Oh (2017)

improvement initiatives generate spillover effects, facilitating their dissemination to other plants. Production workers in overseas plants effectively perform skill-saving work, and labor–management relations are generally congenial. More than 90% of the key positions in charge of daily operations are held by locally recruited managers, and they are managed under a local personnel system. Although the nominal localization rate is high for parts procurement and delivery, the actual rate is significantly lower due to the practice of follow sourcing.

Figure 6.2 summarizes how Hyundai Motor's agile production system has been successfully transferred to overseas plants in green-field locations. The parts of the original system that have been 'applied' include flexible production technology based on the model factory concept, skill-saving work organization, and having Korean parts makers responsible for supplying parts. On the other hand, the HR system for managers at overseas plants demonstrates some characteristics of 'adaptation.'

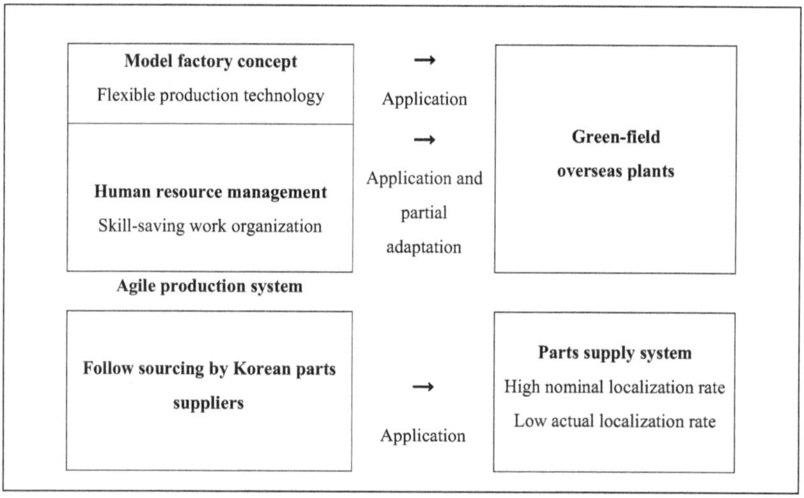

Fig. 6.2 The transfer of Hyundai Motor's production system to overseas plants (*Source* Author's own creation)

6.3.5 *Comparison Between Hyundai Motor and Japanese Automakers*

This section compares the overseas expansion of Hyundai Motor's production with that of Japanese automakers. In line with the progress of globalization progresses, Japanese companies tend to make the organizational capabilities of their headquarters a central part of global expansion. However, there are significant differences from Hyundai Motor Group when it comes to the way in which overseas plants are run.

First, while Hyundai Motor Group has an owner or *chongsu*-eccentric decision-making structure, Japanese automakers have a more decentralized and slower decision-making structure. This difference manifests in the process of transplanting production facilities overseas. While Hyundai Motor engaged in bold and aggressive independent investment to establish overseas plants that mainly produced compact vehicles, Japanese automakers have adopted the more gradual and prudent approach of joint ventures to develop local plants that mainly focus on specialized segment vehicles under a regional division of labor. Japanese automakers initially entered the market through joint ventures with local automakers, then gradually switched to independent investment. As a result there are many cases where overseas plants are also located in brownfield areas (Jo & Jeong, 2016a).

For example, while the Hyundai Motor Group made the decision to build plants in Slovakia (2004) and the Czech Republic (2006) with an annual capacity of 300,000 units each, Japanese automakers established plants in Europe from the late 1980s to the early 1990s. The Japanese companies started with joint ventures in Spain and England on the periphery of Europe, and gradually moved inward to the center of the continent (Fujimoto et al., 1994). The initial production capacity was 100,000 KD assembly units per year, which slowly grew over time in line with a higher degree of localization, globally standardized mass-produced models.

Second, while Hyundai Motor Group built overseas plants based on the concept of globally mass-produced vehicle products and model factories, which entail a high level of standardization. In contrast, Japanese automakers have experimented with a variety of methods.

Kim (2015) classifies Hyundai Motor's product development strategy as 'home country localization.' In other words, while the product development function remained concentrated in the Namyang R&D Center in Korea, the specific needs of overseas customers were incorporated into product development to produce specialized regional models differentiated by design, size, and interior and exterior parts. However, the product development strategies of Japanese automakers differ from company to company. Some also follow the principles of home country localization, while others such as Honda have adopted an 'overseas country localization' model where there is greater localization in product development.

In the relationship between the head office and overseas plants, Japanese automakers vary in the way organizational capabilities are transferred to such plants, and this depends on the global strategy of the head office (Nakayama, 2003). For example, Honda not only designates a parent plant for each vehicle type, but also keeps all data and expertise related to the transfer of organizational capabilities to overseas plants concentrated in the head office's technology research center. Toyota designates parent plants in each region, regardless of vehicle type, and then the parent plant is responsible for the transfer of organizational capabilities to new local plants, and supervises this process. Nissan does not have a parent factory concept, but the domestic engineering department oversees new car development in overseas plants and manages the transfer of organizational capabilities by selecting necessary staff from each domestic plant.

The Hyundai Motor Group is similar to some Japanese automakers in that the head office centrally manages overseas plants through the model factory concept. This appears to be comparable to Nissan in that one stage of pilot car production (P2) is carried out in overseas plants, while the headquarters in the home country oversees new car development (Oh, 2017). However, this process is more headquarters-centered in Hyundai Motor as the facilities to be installed in overseas plants are produced and tested in Korea before being transported to the local plant.

Finally, the relationship with parts suppliers is also different. While Hyundai Motor's overseas plants have developed a vertical and closed relationship revolving around collaborative Korean parts makers in charge of follow sourcing, the overseas plants of Japanese automakers have a more horizontal and open relationship by utilizing the development of parts and sourcing from both Japanese and local parts makers. Accordingly, the Japanese firms have a higher degree of localization in parts sourcing (Jo & Jeong, 2016a).

As summarized in Table 6.6, Japanese automakers also have a strong tendency to involve the headquarters in the operation of overseas plants, but their degree of involvement is weaker than Hyundai Motor Group. This contrast appears to stem from differences in corporate governance. Although the head offices of Japanese automakers have a large degree of involvement in the development of new models and preparation for mass production, the role of overseas plants has gradually increased over time. In particular, overseas plants play a strong role in the development and procurement of parts. Furthermore, compared to Hyundai Motor, Japanese automakers can be regarded as having made considerable progress in transferring organizational capabilities to overseas plants when it comes to everyday operations and human resource management.

6.4 Conclusion

The overseas plant construction of Hyundai Motor has been to transfer the agile production system set-up in Korea to green-field areas. In this process, the company applied the model factory concept with flexible production technology and skill-saving work organization to overseas plants. Korean parts makers that entered overseas markets together with Hyundai Motor were also able to localize their parts supply systems. However, the actual localization is low as parts are sourced to overseas plants by combining knockdown exports and local production, while the development of parts remains in Korea.

Applying Abo's (1994) concept, Hyundai Motor can be regarded as having succeeded in overseas production by 'applying' the company's production system. In Chapter 1 we described the technical conditions of 'machine technology' that made it relatively easy for Hyundai Motor to catch up with advanced companies as a latecomer, and it is this same technology that facilitated the rapid transfer of the company's production system to overseas production (Veblen, 1915). This method is distinct from the way in which Toyota and other Japanese automakers have actively promoted localization and gradually 'adapted' to new overseas locations.

In recent times, Hyundai Motor Group has made efforts to increase the localization rate by establishing a regional headquarters that integrates the production and sales functions of overseas plants (Interviews with a Hyundai Motor's executive in 2021). However, considering the path dependence of the agile production system, this is not expected to lead to a significant change in the operations of overseas plants.

Table 6.6 Transferability of organizational capabilities in Japanese and Hyundai Motor Group's overseas automotive plants

		Japanese automakers	*Hyundai Motor Group*
Decision-making process		Mid 1980s–Mid 1990s Passive entry strategy due to trade friction and highly valued yen Joint ventures with local layers, high level of relative autonomy	Late 1990s–Mid 2000s Active entry strategy to secure the market Sole proprietorship (100% stake), little relative autonomy
Plant construction		Parent factory-subsidiary relationship	Model factory concept, test run of facilities in Korea
Non-routine plant operations	Product development	Led by the head office's research institute, Active participation from local design centers in the development of regionally specialized vehicles	Led by the head office's research institute, Less participation from local design centers in the development of regionally specialized vehicles
	Preparation for mass production	Led by HQ, Active participation of local overseas plants in pilot production	Led by HQ, Less participation from local overseas plants in pilot production
	Parts development	Led by local parts suppliers, High localization rate	Led by Korean parts suppliers responsible for follow sourcing, Low localization rate
Routine plant operations	Production operation	Overseas plants lead improvement activities, Active encouragement for workers to become multi-skilled	Gradual progress in improvement activities in overseas plants, Less encouragement for workers to become multi-skilled
	Human resource management	Localized human resource management, Collaborative labor–management relations	Localized human resource management, Conformist labor–management relations

Source Author's own creation

References

Abo, T. (1994). *Hybrid Factory: The Japanese Production System in the United States*. Oxford University Press.

Black, J. (1997). *Oxford Dictionary of Economics*. Oxford University Press.

Fujimoto, T. (2006). Architecture-Based Comparative Advantage in Japan and Asia. In K. Ohno & T. Fujimoto (Eds.), *Industrialization of Developing Countries: Analysis by Japanese Economics* (pp. 1–10). National Graduate Institute of Policy Studies.

Fujimoto, T., Nishiguchi, T., & Sei, S. (1994). The Strategy and Structure of Japanese Automobile Manufacturers in Europe. In M. Mason & D. Encarnation (Eds.), *Does Ownership Matter? Japanese Multinationals in Europe* (pp. 367–406). Clarendon Press.

Guo, S., & Upathamwaranon, P. (2008). *The Internationalization Process of Toyota in Europe: From the Perspectives of Entry Mode and Network Structure* (Working Paper, School of Sustainable Development of Society and Technology, Mälardalen University).

Jo, H. J., & Jeong, J. H. (2016a). Whether or not the Overseas Transferability of Hyundai Production System? A Focus of the Supplier Relations of Hyundai Motor Manufacturing Czech (HMMC). *Journal of Korean Social Trend and Perspective, 96*, 48–80. (In Korean).

Jo, H. J., & Jeong, J. H. (2016b). Human Resource Management of Hyundai Motor Group's European Subsidiaries: The Dynamics of 'Application' and 'Adaptation.' *Korean Journal of Sociology, 50*(5), 79–106. (In Korean).

Jo, H. J., & Jeong, J. H. (2017). Global Operations of the Organizational Capability at Hyundai Motor Company: A Case Study of A Transplant in Europe. *Korean Journal of Labor Studies, 23*(3), 281–313. (In Korean).

Jo, H. J., & You, J.-S. (2011). Transferring Production Systems: An Institutionalist Account of Hyundai Motor Company in the United States. *Journal of East Asian Studies, 11*(1), 41–73.

Kim, C., & Oh, J. (2017). Analysis on Oversea Productions of the Korean TNCs: Focusing on Suppliers that Accompanied Hyundai Motor Group in Central Europe. *Korean Journal of Sociology, 51*(1), 129–154. (In Korean).

Kim, H.-J. (2015). *The Localization of Product Development in the Case of Denso: The Change of Headquarter Organization and Knowledge Link*. Yoobigak Press. (In Japanese).

Nakayama, K. (2003). Technological Support by Mother Factory System of Japanese Automobile Manufacturers. *Meijo Review, 3*(4), 35–58. (In Japanese).

Nobeoka, K. (2007). The Stack of Organizational Capability: Technology that Cannot Be Imitated. *Organizational Science, 40*(4), 4–14. (In Japanese).

Oh, J. (2017). Comparative Study of Vehicle Development Systems for Emerging Markets: Focusing on the Cases of Nissan and Hyundai Motors. *Musashi University Journal*, *65*(1), 49–58. (In Japanese).
Veblen, T. (1915). *Imperial Germany and the Industrial Revolution*. Macmillan.

PART III

Reflections and Prospects

CHAPTER 7

Conclusions—Theoretical and Empirical Reflections and Future Prospects

In this book, we have attempted to pry open the 'black box' of compressed growth for Hyundai Motor, which had stagnated on the brink of the 'middle-ranked carmaker's trap' in the 1990s, since the 1997 Asian Financial Crisis. Hyundai Motor then took off as a major finished car maker in the 2010s as a result of *chaebol* corporate governance and overhauling its production system. In this chapter, we summarize the theoretical and empirical discussions surrounding Hyundai Motor's production system, reflect on the two sides of this system, and discuss the future prospects for Hyundai Motor in the face of rapid transformation in the automobile industry with the advent of electric and autonomous vehicles.

7.1 Theoretical Contributions

In this book, we have explained the driving force behind Hyundai Motor's compressed growth as an agile production system that is distinct from Toyota's. This agile production system is characterized by the ability of engineers to solve problems in an improvisational manner, which is supported by leadership that grants engineers a certain degree of autonomy, flexible work organization and low-skilled shop-floor workers involving flexible automation, skill-saving and segmented labor, and extended quasi-vertical suppliers relationships. We call this a type of

H. J. Jo et al., *Agile Against Lean*,
https://doi.org/10.1007/978-981-99-2042-6_7

'authoritarian experimentalism' in which Hyundai Motor is still a pre-Chandlerian *chaebol* with no separation between ownership and management and, nevertheless, it has simultaneously carried out rather different post-Chandlerian practices in nature through utilizing typical Chandlerian managerial resources, leading to granting autonomy to a specific group (engineers), despite a little limited in scope, and modularizing parts supply relationships.

7.1.1 *The Catch-Up Process*

Veblen (1915) and Gerschenkron (1962) have somewhat different views on the catch-up process. The former discusses the relative ease of catching up to advanced players and raises the possibility of a market solution through the introduction of imported technology, while the latter believes catching up is difficult without intervention by a third party such as a government or other non-market institutional apparatus. State-of-the-art machines that combine technology and skills are now widely used due to the lower prices of capital goods. The relative price of capital goods to labor has steadily declined worldwide since the 1980s, coinciding with low-interest rates (Karabarbounis & Neiman, 2013). However, this trend may soon be reversed due to the US-China struggle for supremacy and inflationary pressure.

Since the 1990s, interface design between product architectures has accelerated modularization, which has in turn facilitated an international division of labor (Shintaku et al., 2006). By achieving the ability to compete on costs through large-scale investment in modularized products, latecomers can now catch up to developed countries in a way that was not possible in the past. Traditionally, new products were developed and produced in developed countries before production facilities were transferred to low-cost developing nations as the technology matured, and it took a considerable period of time for latecomers to catch up (Vernon, 1966). However, since the 1990s, modularization has accelerated the geographic transfer of production, and the lag time for catching up has rapidly declined. This implies that the relative ease of technology transfer due to standardization based on codified knowledge has allowed latecomers to catch up in a more leapfrogging way, involving the latest technology's coexistence between leaders and latecomers at the spatial level unlike before.

In particular, recent automated facilities and tools can be programmed on microchips, making expertise and skills 'encapsulated.' If these latest facilities and tools could be purchased, although the expertise and skills built into the microchips are found to be a black box, there is no big problem in using it. Thus, catching up may be more likely to be easier than in the past. However, technical learning is still required to absorb and acquire the tacit knowledge behind technology embodied in the black box. The reduction in catch-up time lag does not mean that the economic gains of developed countries have fallen, as companies in developed countries tend to focus on developing technology-intensive integrated architecture products with a high level of added value (Shintaku et al., 2006).

Given that technological development and the effects of globalization should be considered in this discussion, we believe it is necessary to connect the respective views of Veblen (1915) and Gerschenkron (1962) rather than exclude them. Even though learning technology is still a difficult process, the degree of learning now varies depending on the international division of labor, technological development patterns, and the stage of the catch-up process, and is contingent on the social and institutional context. In this book, we have paid particular attention to Hyundai Motor's catch-up 2.0 period. By the time of 1997 Asian Financial Crisis, Hyundai Motor had already completed the 'catch-up 1.0' period of developing its own engines and models. The investment-based growth model through the entrepreneurial dynamism of *chaebols*, which Amsden (1989) and Aghion et al. (2021) refer to, is a good example of this, as is the capability building approach based on the reverse life product cycle (Kim, 1997), which assumes that technological development occurs sequentially and over a long period of time.

However, this discussion is quite different from the catch-up theory of Gerschenkron (1962) which states that catching up is only possible when the latecomer attempts something that differs from the predecessor. Lee (2013)'s technology life cycle theory and Aghion et al. (2021)'s 'innovation-based' growth model both provide a discussion of 'catch-up 2.0' in which middle-income countries that have succeeded in an initial catch-up period manage to avoid falling into the middle-income trap by using opportunity windows such as technological change or policy reform. This suggests that the economic and institutional arrangements that facilitate catching up are different depending on the catch-up stage. Although Lee's (2013) technology life cycle theory has great applicability in explaining the catch-up of short-lived technologies or technologies with a short product life cycle such as information technology, it is less relevant in industries where gradual innovation is still important, such as the automobile industry. Lee and Lim (2001) described Hyundai Motor's engine development as an example of 'path-skipping' catch-up, but failed to pinpoint a specific learning mechanism or channel that made this possible. Aghion et al. (2021)'s innovation-based growth theory is also a quantitative analysis, simply emphasizing the effect of open and competition-encouraging policies without specifically explaining how the Korean economy avoided falling into the middle-income trap. Unlike these recent studies, we analyzed the 'catch-up 2.0' process of Hyundai Motor and the mechanisms through which this took place since the 1997 Asian Financial Crisis. We believe that Hyundai Motor has developed into a top five global automaker through entrepreneurial investment and internal innovation that embodies both the positive and negative sides of the *chaebol* system. The company was able to utilize windows of opportunity such as the global trend toward modularization and domestic conditions in Korea including the takeover of Kia Motors after the financial crisis. In this sense, this book goes beyond existing studies such as Amsden (1989), Lee (2013), and Aghion et al. (2021) by focusing on the catch-up 2.0 period, where economic and institutional conditions were markedly different from the initial catch-up period. We also critically examine the dual nature of the *chaebol* system instead of simply focusing on the company's success story.

7.1.2 *The* Chaebol *System*

In Chapter 1, we looked at the five types of business organizations in relation to catch-up in economic growth, mainly through latecomers such as the United States, Japan, Germany, Italy, and Korea. These include Marshallian SMEs, Chandlerian big corporations, Neo-Schumpeterian evolutionary companies, and post-Chandlerian network companies involving quasi-vertical integration, modular relational contracts, and pragmatic and experimental collaborations between economic agents, as well as Korean-style *chaebol* business groups. These types of enterprises sit at the core of an ecosystem that drives economic development in a market economy within a specific historical and institutional context (Teece, 2017). The dynamism of national economies is inextricably linked to the dynamism of enterprises that operate in these different business worlds (Chandler, 2001: 5).

As noted in previous chapters, Fujimoto (2006) pointed out that the concentration of rapid decision-making is a core competence of Korean *chaebols*. *Chaebols* can be regarded as Chandlerian corporations built on the foundation of a pre-Chandlerian ownership structure, with relatively weak external checks and supervision. As pointed out by Amsden (1989), the entrepreneurial dynamism of *chaebols* has led to a concentration of swift investment decisions in a positive direction thanks to the state's discipline. However, if such investments fail, the *chaebol's* resource allocation and investment decisions may be made as a means of private extortion to socialize losses or maintain the identity of the family business thanks to the myth of being 'too-big-to-fail.'

In the post-financial crisis period, the state's discipline for *chaebol* no longer worked. The market discipline of external investors is also concerned with maximizing short-term shareholder value rather than establishing a rational and fair trade basis for the market. There is little interest in improving the *chaebol's* governance structure as long as the group continues to grow and generate cash flow. In this context, it is important to explain how the dual nature of the *chaebols* emerges.

We have attempted to highlight the dual nature of the *chaebol* system and suggested that this duality could be called 'authoritarian experimentalism.' This is sort of an oxymoron, in the same vein as Adler and Cole (1993) referring to NUMMI as 'democratic Taylorism' or 'learning bureaucracy.' In this expression, 'authoritarian' refers to the vertical governance structure of the *chaebol*, where the owner or *chongsu* exercise full control over the business group called the *chongsu*-holder value. On the other hand, 'experimentalism' refers to the entrepreneurial risk-taking behavior of *chaebol* owners or executives to capture the value of real options that may pursue long-term growth or returns above the market average and to collective learning and its economic achievements accompanied by trial and error. Some examples of this include giving engineers a certain degree of autonomy and involving stakeholders such as affiliates of the business group and outside parts suppliers to improve product quality. In other words, growth is embedded in developmentalism, which entails investment and coordination with risk-taking, collective learning, and experimentation through the commitment of various stakeholders, as well as the resulting financial performance. This implies concentration and exclusion, and the possibility of committing a Type 2 error of actually judging an irrational investment as a rational one. Accordingly, the impressive growth of Hyundai Motor also has a dark side.

Since the 1997 Asian Financial Crisis, Hyundai Motor has pursued the modularization of automobile parts to reduce costs and risks and improve flexibility and also engaged in widespread outsourcing by leveraging the *chaebol* system of extended quasi-vertical integration. It is to extend beyond direct affiliates of the business group to include external subcontractors by exploiting the company's monopsony on demand. As Hyundai Motor slimmed down its parts sourcing system, the ecosystem in the automotive parts industry risked falling into the 'modularization trap' (Whittaker et al., 2020: 77) where excessive dependence on a specific automaker lowers resilience and the concentration of economic power leads to antitrust abuses and unequal bargaining power in negotiations (Mosca, 2018).

7.1.3 The Automotive Production System

The lean production system is the current industry standard in automobile production. Described by Womack et al. (1990) as 'the one best way,' the lean production system explains why Toyota has maintained the greatest competitive advantage in the global market since the 1980s. This is why the lean production system has been benchmarked as a best practice for not only automobile manufacturers, but also services enterprises (Janoski & Lepadatu, 2021). The literature on automobile production systems shows diversity through the concept of 'productive models' such as Boyer and Freyssenet (2002), while criticizing Womack et al.'s (1990) argument of 'one best way.' Janoski and Lepadatu (2021) also note the variation among lean production systems.

Hyundai Motor maintained a close relationship with Ford and Mitsubishi during the initial catch-up period and attempted to embrace the lean production system from the mid-1980s. At first glance, the Hyundai Motor's production system can be understood as a combination or hybrid mass production system based on Taylorism and Fordism, and Toyota's lean production system. The lean production system is an evolved, but different form of mass production. The former has achieved integration between management and shop-floor workers, avoiding separation between concept and execution, and differs from the latter in that production is streamlined through a just-in-time (JIT) sourcing system between the final car assembler and parts suppliers.

We believe that Hyundai Motor's transformation of both the existing mass production and lean production system occurred during the catch-up 2.0 period. We call the resulting system an 'agile' production system. In terms of the corporate organization archetypes discussed above, Toyota has attributes of a Chandlerian corporation, as it basically inherits the mass production system of producing standardized products in volume through an assembly line. Moreover, although Toyota's system of *keiretsu* cooperation between companies can be interpreted as 'experimental' cooperation (Herrigel & Sabel, 1999), the company can also be viewed as more of a neo-Schumpetarian evolutionary big firm that secures an ongoing competitive advantage through cumulative improvements in manufacturing capabilities (Fujimoto, 1999).

Building cumulative capabilities requires a lot of time and effort, and it is difficult for latecomers to fully absorb the best practices of first movers. This makes it almost impossible for latecomers to catch up using this strategy. We believe that Hyundai Motor's 'catch-up 2.0' went beyond simply benchmarking Toyota, and involved absorbing the practices of post-Chandlerian automotive firms such as pragmatic, experimental, or improvisational learning practices. This was achieved by granting autonomy to engineers and building a system of modularization based on extended quasi-vertical integration. Post-Chandlerian practices are generally defined as pragmatic experimentalism and modularization. The former is aimed at integrating concepts and execution among stakeholders through mutual participation and interactive definition and redefinition of their roles and identities, creating dynamic capabilities in a non-hierarchical manner (Herrigel, 2020). The latter refers to open coordination and cooperation between vertically quasi-integrated and specialized firms through standardized technology interfaces (Sturgeon, 2002).

Under a pre-Chandlerian control structure with no separation between ownership and management, Hyundai Motor introduced experimental and improvised elements in the process of car development during the catch-up 2.0 period. In Hyundai Motor, routine does not precede capability unlike in Toyota. Instead, the former has been backfilled and updated through the improvisational capability to execute projects based on the delegation of authority to engineers, and thus the latter has been built in a non-sequential fashion rather than in a conventional cumulative and linear manner as in Toyota (Gong et al., 2006). For example, Hyundai Motor built a large-scale pilot center within the R&D center to identify and solve issues in product development through the participation of engineers, parts makers, and production workers. In this way, the company builds up problem-solving capabilities in new car development through simultaneous engineering, benchmarking, and joint deliberations. We have called this process 'learning by improvisation.' This is our attempt to pinpoint a mechanism for integrating skills within an organization through engineer-led skill formation. It goes beyond Hattori's (2005) hypothesis of 'assembly-type industrialization' where products are exported abroad by simply putting together imported materials and parts, and his discussion of a divide between technology and skill in Korea (Hattori, 1986). However, this mechanism differs from Toyota in that it excludes shop-floor labor and seeks to make up for the role of shop-floor workers by placing an excessive burden on skilled workers and engineers.

Hyundai Motor has also modularized production in a closed fashion through 'extended quasi-vertical integration,' where the unique *chaebol* system of quasi-vertical integration has been extended to external partners. This differs from MacDuffie's (2013) comparative study on parts modularization between Ford and Hyundai Motor. This strategy extends the *chaebol's* distinct project execution capabilities to parts suppliers by sharing resources and coordinating investment, and is characterized by both 'opportunity' and 'exploitation.' Although there is risk sharing in the form of providing stable markets and growth opportunities to external parts makers along with technological advice, investment support, and a stable sales channel, there are also exploitative elements such as technology theft, periodic unit price cuts, and ambiguous contract clauses that blur the boundaries between the companies.

In order to capture the post-Chandlerian nature of Hyundai Motor's structure that appeared during the catch-up 2.0 period, we refer to this as an 'agile' production system. The term 'agility' first emerged in 1991 in a report on new foundations of competition by the Iacocca Institute at Lehigh University (Nagel, 1991). This report suggested that major issues in manufacturing are continuous change, agile responding, quality improvement, and social responsibility. Gunasekaran (1998) describes agile manufacturing as the ability to create a competitive advantage in a competitive environment of constant and unpredictable change by responding quickly to changing markets driven by product and service design in response to customer needs. In contrast, leanness refers to synchronizing value streams by maintaining a certain level of production schedules through continuous process improvement to eliminate waste within the production system (Womack et al., 1990). The leanness focuses on the productive use of limited resources to counter competitive pressures, whereas the agileness emphasizes taking advantage of beneficial opportunities as a strategy and overall response to constant change.

In lean production, zero inventory is maintained to eliminate all non-value-added processes, while in agile production, the inventory level may differ according to changing end user requirements. Agile production gives the company the ability to quickly reconfigure the production process to respond to market demand, but is less concerned with reducing waste during this process. In contrast, lean production focuses on minimizing the processes and time wasted in converting to a new product. Since this requires prior warning of production schedules and product changes, rapid process reconfiguration is less prominent. Lean production systems respond to volatility in the speed of production by streamlining the supply chain to maintain stable demand and plan for the future using market knowledge and data, while agile production system is more concerned with building a supply chain that can deliver a variety of products and cope with end user fluctuations (Naylor et al., 1999).

Although there are subtle differences between lean and agile production systems, the two have many things in common. Agile production allows for greater manager discretion and flexibility in resource allocation and strategy formation to respond to market uncertainties. As such, the theoretical attempt to define Hyundai Motor's structure as an agile production system corresponds to Boyer and Freyssenet's (2002) discussion on diversity in current automobile productive models, and is somewhat different from Janoski and Lepadatu's (2021) discussion on variation in lean production systems.

7.2 Empirical Findings

So far, we have highlighted the differences with prior literature from a mainly theoretical standpoint. And then, in what follows, we will briefly summarize the empirical work in the previous chapters where Hyundai Motor's production system has been conceptualized as the authoritarian experimentalism and agile production system. Moreover, the main characteristics of Hyundai Motor's production system will be recapped through a comparison with Toyota.

7.2.1 Skill-Building in Engineers

Although Hyundai Motor's top management presented the abstract goal of producing high-quality vehicles under the banner of 'quality management' in the wake of the 1997 Asian Financial Crisis, the company has developed dynamic organizational capabilities by quickly solving the problems it faced through trial and error, improvisational learning, and cooperation and deliberation with other stakeholders. This demonstrates one aspect of experimentalism, where problems are solved through the mutual participation of relevant stakeholders (Herrigel, 2020).

Even in a business organization with a vertical hierarchy such as a *chaebol*, this was made possible by the company's implicit organizational practice of delegating authority to engineers and not paying too much attention to the mistakes that occurred in this process. The opinion of engineers on the shop-floor is valued as much as that of financial managers. This practice provides a space for autonomy to encourage experimental solutions within the organization, and the *chaebol* has the sufficient market power to allow different stakeholders to participate and discuss the development and mass production of new vehicles.

As discussed Chapter 3, the dedicated pilot center built at Namyang R&D Center in 2004 served to shorten the development period through design and mass production quality tests, thereby reducing the time it took to set up overseas plants and speeding up assembly work through time and motion studies and thus rendering engineers to experience and learning. This has resulted in a shorter development period for the company. In particular, the pilot center has functioned as a place for integrating stakeholders such as shop-floor workers and parts suppliers into the organization. The fact that the pilot center is located in the research center shows the space that provides convenience to engineers, although with somewhat a physical and psychological distance from the shop-floor. In addition, there has been a gradual division of roles between manufacturing design technology and process technology engineers since the opening of the pilot center, with the former now playing a stronger role in the development of new vehicles.

The pilot center exemplifies the fact that the intensive problem-solving ability of Hyundai's engineers has been maximized in a 'leapfrogging' way rather than sequentially, which differs from Toyota. At Toyota, there are no fixed assumptions about intra- and inter-organizational relationships, and the organization's starting point is the provisional stipulation of goals. In this respect, it is experimental. In contrast, different stakeholders at Hyundai Motor are *de facto* integrated into the business through the *chaebol's* market power. In addition, engineers very intensively build dynamic skills and organizational capabilities across existing boundaries through the project-based CFT system at the pilot center called 'learning by improvisation,' as described in Chapter 3. In particular, since all new car developments targeting domestic and overseas markets at Hyundai Motor were actually concentrated at the Namyang Research and Development Center, the engineer's learning curve has increased very steeply there. In this respect, Hyundai Motor's system can also be viewed as experimental.

7.2.2 Work Organization and Shop-Floor Workers

Hyundai Motor's production system has been intertwined with labor–management relations and the production philosophy of engineers since the 1990s. Unlike the Toyota production system which leverages on-site skills, Hyundai Motor has moved toward a system that saves on skills. The technological conditions that made this possible are IT-based automation and informatization. This is similar to the way in which Fordism developed in the early twentieth century through the '3S's' (standardization, specialization, and simplification) and conveyor belts. To reduce costs in inventory management, Toyota has pursued the just-in-time (JIT) system that produces only as much as necessary, while Hyundai Motor has created just-in-sequence (JIS), a system that delivers modularized parts in sequence.[1] While Toyota aims for zero defects on the work line and implements quality controls, Hyundai Motor saves on costs and

[1] The JIS system means that Hyundai Motor *de facto* can produce a mixed range of products on its own assembly line because its suppliers deliver parts and components in accordance with Hyundai Motor's sequential production orders. Hyundai Motor has been able to reduce costs through such inter-firm relationships. In contrast, Toyota has been able to reduce costs by producing multiple models at the same time through the JIT system.

time by using information technology to overcome the oversizing of inspections inherent in Fordism and improve the inspection process by identifying potential problems in advance, thanks to committed pre-work by engineers.

Hyundai Motor's production system is Tayloristic in that it involves a division of labor between engineers and workers in which concept and execution are strictly separated, but also Fordist in that it employs economies of scale and automation to replace labor. However, Hyundai Motor's model somewhat differs from a typical American Fordist production system in that engineers not only provide fixed standards of tasks, but also play some supplementary role to shop-floor workers to integrate the tasks between the shop-floor and the engineer, with the latter's no existence in the typical American system. In this way in which both the IT-based automation and informatization and the committed engineer's pre-work has been exploited, Hyundai Motor's production system overcomes the inefficiencies of the American Fordist production system without relying on the skills of shop-floor workers.

Immediately after the Asian Financial Crisis in 1997, Hyundai Motor laid the groundwork for financial recovery by reducing labor costs through mass layoffs of workers. After this incident, the union sought to thwart the company's attempt to flexibly utilize the labor force and manage human resources in a performance-oriented manner due to deep distrust. In addition, flexibility in the labor market due to neoliberal reforms in the wake of the financial crisis ultimately led to the regular workers' union signing a full-employment protection agreement with the company in 2000, which amounted to tacitly accepting the exploitation of non-regular workers.

Although Hyundai Motor has not carried out aggressive facility investment on the same level as before the financial crisis, a comparative advantage has been achieved based on economies of scale by maximizing operating time. This requires long working hours, which makes it difficult to establish a system that enables shop-floor workers to build up a high level of skill. However, the operating hours of Hyundai Motor's facilities have gradually decreased since the mid-2000s due to criticism over long working hours and intervention from labor unions.[2]

[2] In addition to increased wages, each union member sought to increase their own total pay through long working hours in line with the company's long operating hours. Paradoxically, this also served as an obstacle to reducing excessive working hours.

7.2.3 *Inter-Firm Relationships*

Hyundai Mobis is the core modularization company within the Hyundai Motor Group. Hyundai Mobis began large-scale production of modular parts in the 2000s by reorganizing affiliates and merging with or acquiring external companies. Hyundai Mobis came to hold a virtual monopoly over the production of three major modules (cockpit, chassis, and front-ends), and sourcing of modular parts from non-affiliates has been limited. Hyundai Mobis oversees the parts supply network on behalf of Hyundai Motor, and designs modular parts through close cooperation with the finished manufacturer. This has enabled Hyundai Motor to achieve a multi-layered parts supply structure and hierarchical control.

Since the purpose of modularization is to outsource and break up the assembly process before supplying parts to the main line in bulk, this strategy serves to reduce the increased complexity in the main line from an increasing number of models. This system is efficient when there is a wage gap between finished makers and parts suppliers. Hyundai Motor's modularization includes most of the body and final assembly parts. By placing multiple orders for these parts from a small number of affiliates or external parts makers, the company effectively concentrates the work flow around the main assembly line and increases the automation rate (Oh, 2022). In this way, Hyundai Motor's production system can be characterized by modularization, automation, and simplification, just like the Fordist one. With Hyundai Motor's benchmarking Volkswagen, modularization has become a core component of Hyundai Motor's competitive advantage through extended quasi-vertical integration.

7.2.4 *Overseas Transplants*

A firm requires a combination of technical, organizational, and spatial fixes in order to grow. Hyundai Motor has chosen to leave the technical and organizational fixes in Korea and use overseas locations as a spatial fix. The globalization strategy bypasses the local employment system and attempts to transfer the Korean production system, thereby achieving results that are almost indistinguishable from those of the parent plant in Korea. From the time Hyundai Motor entered emerging markets, the company has focused on minimizing manufacturing costs by building mass production plants and maintaining a high utilization rate. These large-scale transplants have helped suppliers responsible for follow sourcing to enter the market together, increased the local procurement rate, and established a system for the smooth sourcing of modular parts. In short, Hyundai has transferred its system of extended quasi-vertical integration overseas. With Hyundai Mobis at the forefront of follow sourcing, pure local suppliers have been largely shut out of the parts supply.

Hyundai Motor established the concept of a model factory, and the R&D center in Namyang led the construction of overseas local plants. This is a strategy to transfer and maximize knowledge from Korea instead of utilizing local knowledge. In this process, engineers at the pilot center have been able to share different experiences as they go through trial and error and accumulate expertise. In particular, the frequency of interactions between the Korean headquarters and the local area increased as Hyundai's overseas expansion progressed rapidly, providing an opportunity for engineers to build organizational capabilities at the pilot center because, as mentioned before, virtually all new car development and overseas plant set-up took intensively place at the Namyang R&D Center. Hyundai Motor has successfully transferred its agile production system to green-field locations, only minimally adapting to institutional conditions overseas. Using Abo's concept (1994), Hyundai Motor can be viewed as having succeeded in overseas production by 'applying' the company's production system. The same 'machine-technology' that helped Hyundai Motor to catch up to advanced players as a latecomer also facilitated the rapid transfer of the production system to overseas locations.

7.2.5 Comparison Between Hyundai Motor and Toyota's Production Systems

Hyundai Motor's production system appears to be too different to be viewed as a variant of the lean production system. It is true that the lean production system has had a great influence on Hyundai Motor, but the company created its own production system during the second catch-up period. Hyundai Motor's production system incorporates both engineer-led improvisation and pragmatic experimentalism that largely excludes shop-floor labor. This system exists under the *chaebol* corporate governance structure in which the *chongsu* exerts influence that is entrepreneurial in some cases but discretionary in others. In this sense,

Hyundai Motor's production structure can be summarized as a combination of 'engineer-led automation' and 'standardization of labor processes' (Jo & You, 2011).

Table 7.1 compares the main characteristics of Hyundai Motor and Toyota's production systems. Toyota's system is rooted in Taylor-Fordism but also adds a number of variations in that it attempted to integrate concept and execution. In contrast, Hyundai Motor's system also evolved from Taylor-Fordism, but is a transformation and remake of the American agile production system. This system maintains the separation between concept and execution (Taylorism), but also uses machines to substitute for labor, which is strongly oriented toward a technocratic solution (Fordism). Hyundai Motor has adopted new production technologies to enhance flexibility since the 1990s, but the existing Taylor-Fordist work structure, which relies on the segmented division of labor and simple repetitive unskilled tasks, has remained in place. This is quite different from a lean production system that attempts to integrate concept and execution through shop-floor labor participation. However, Hyundai Motor's production system did benchmark certain elements of the lean production system in its own way. For example, Hyundai Motor has partially solved the problem of integrating development and mass production, an issue which the lean production system traditionally struggles with.

Hyundai Motor's work process is based on a 'push' system in which parts are sequentially supplied in the process order. However, by benchmarking the JIT system where the flow of the entire work process is synchronized across company boundaries, Hyundai Motor has devised a new just-in-sequence (JIS) system in which external parts suppliers deliver components based on production orders. This system enables Hyundai Motor to produce a mix of models on the same line and reduce costs, but this cost burden is passed on to external suppliers. In this regard, although the parts supply structure appears to have become more multilayered and the captive transaction structure has been somewhat mitigated in the wake of restructuring among Korean automakers in the 1990s and the full-scale modularization and platform integration of Hyundai Motor Group in the 2000s, as the supply of core modular parts has mainly been centered on core affiliates, Hyundai Motor Group's actual control over external partners has strengthened.

By benchmarking Germany's modularization and pilot center, Hyundai Motor was able to escape from the 'middle-ranked carmakers trap' that

Table 7.1 Comparison of Hyundai motor and Toyota's production systems

	Hyundai motor	*Toyota*
Governance structure	*Chaebol* (hierarchical structure; *chongsu*-holder values)	*Keiretsu* (dual structure)
Profit strategy	Flexible scale and diversity	Constant cost reductions
Product strategy	Moving from low-mid to high-end; platform integration	A variety of product portfolios; platform integration
Labor relations	Confrontational, segmented dual structure	Cooperative, integrated (career ladder)
Work organization	Separation between concept and execution (fragmented)	Integration between concept and execution (integrated)
Process flow	Push system based on informatization and automation; just-in-sequence sourcing system	Pull system based on *Kanban*; just-in-time sourcing system
Product development (between development and mass production)	Engineer-led partial integration (pilot center within the Namyang Research Center)	Integration through shop-floor labor participation (global production center in parent plant)
Capability building (major learning mechanism)	Engineer-led 'learning by improvisation'	'Learning by monitoring' via engineer and shop-floor workers participation
Inter-firm relations	Captive, multi-layered (high modularization)	Captive, cooperative (low modularization)
Overseas transplant strategy	Aggressive; spatial transfer of domestic production system	Gradual; 'learn locally and act globally'
Organizing principles	Authoritarian experimentalism	Democratic experimentalism (Dorf and Sabel, 1998); learning bureaucracy or democratic Taylorism (Adler and Cole, 1993)

Source Author's own creation

it faced since the 1990s. In order to pursue a profit strategy of flexible scale and diversity, Hyundai Motor made a bold investment in fixed capital based on flexible automation and informatization in a way that reflected the technology-oriented production philosophy of engineers and hostile labor–management relations. The company has gradually developed a diverse product lineup and further integrated its platform with the acquisition of Kia. In addition, Hyundai Motor has exerted market dominance through the use of carrots and sticks for external subcontractors by leveraging the *chaebol's* monopsony on demand in the business ecosystem. In other words, extended quasi-vertical integration has allowed Hyundai Motor to enjoy external economies of scale through JIS-based modularization.

Hyundai Motor's production system requires consistent sales growth in order to pursue an investment-led strategy of replacing labor and maintaining dominance in long-term relationships with external partners. This means that launching overseas production was an inevitable step for the company. As globalization became more prominent during the post-Cold War era, Hyundai Motor began aggressively expanding overseas production. Hyundai Motor's strategy mainly involved transferring the Korean production system by entering overseas markets together with affiliates and external parts makers through 'follow sourcing' while also taking advantage of low labor costs and production flexibility by favoring green-field investments where there is no need to contend with labor unions.

The dynamic capability building that intensively occurs at Hyundai Motor's pilot center is based on a circular model of 'company-wide goals (quality management) → delegation of authority and granting of autonomy (improvisational learning by engineers) → evaluation and justification (CFT meeting) → formation of new routines.' This is different from prior literature which states that routine precedes capability and organizational capability forms in a cumulative and sequential manner. At Hyundai Motor, this process is more akin to that of a startup, where organizational capability is formed in a non-sequential and 'leapfrogging manner.'

The role of the *chongsu* as the control-holder has been important in Hyundai Motor's successful catch-up. For example, the acquisition of Kia, the implementation of quality management, the construction of the pilot center, and the introduction of modularization were all the results of high-risk decisions made by the head of the *chaebol*. In addition, by giving engineers a certain degree autonomy and running the project-based CFT, it was possible to build organizational capabilities through improvisation due to the entrepreneurial leadership of the head. This leadership creates real option value through bold investment and risk-taking in accordance with changing circumstances, but this entails the possibility of making a Type 2 error due to a lack of external checks and balances. For example, in the mid-to-late 2010s, Hyundai Motor faced an immediate crisis due to excessive investment in China and subsequent geopolitical tensions, which led to a decline in profitability. This discretionary power akin to an 'absolute monarch' could make Hyundai Motor dynamic and highly volatile. In addition, there may exist such circumstances that the head of the *chaebol* is a rational decision-maker akin to an 'enlightened monarch' or a commissioner which empowers competent professional managers akin to a 'constitutional monarch.' These circumstances would be contingent on who leads the *chaebol* group, still due to the weak double veto procedure from outside. As such, the *chaebol* system has a structural advantage in achieving collective efficiency, but also suffers from the instability, arbitrariness, and discretionary judgment of a governance structure based on strict ownership relations. Thus, the bottom line is that Hyundai Motor has accommodated partial empowering and full modularization in an entrepreneurial but authoritarian way, which can be called as an example of 'authoritarian experimentalism.'.

As is well known, the Toyota production system is characterized by JIT-based inventory management, a vertical but collaborative subcontracting system, statistical process management, and value-added engineering. Unlike in Taylorism, Toyota encourages voluntary participation from shop-floor labor and integrates concepts and execution. The institutional basis for this is long-term employment security based on seniority, a merit-based promotion system called 'ranking hierarchy,' and amicable enterprise unions (Aoki, 1988). Toyota's corporate governance structure is called *keiretsu*, and Aoki (1990) sees the Japanese company as jointly owned by the main bank and its employees. The manager coordinates and mediates between these parties. As a result, all collaborators within a *keiretsu* production network are effectively co-owners of the assets under their control, and cooperative subcontracting relationships form a sort of federation of inter-firm relationships in which goals are jointly set between suppliers and customers. Aoki (2010) argues that residual use rights such as human assets are more important than residual control rights such as physical assets or ownership stakes in these production systems.

Toyota gives autonomy to shop-floor labor, and workers become multi-skilled through on-the-job training in teams or 'quality circles.' Employees act as managers in a 'learning by monitoring' system. The JIT system serves to synchronize the flow of the entire manufacturing process, and the company uses a 'pull' system in which subsequent processes take parts from previous processes. The production process must be synchronized within the entire production system across company boundaries in order to achieve flexible, zero-inventory production. The smooth operation of this JIT system requires perfect quality from parts suppliers. Since continual improvement to increase added value takes place between these process flows, the profit strategy of constant cost reductions is effective.

Toyota's internationalization strategy has been cautious and gradual. The company opted for low-risk methods such as exports or licensing in the early stages, then changed to higher-risk methods through joint ventures or overseas plant construction once the local market and networks had grown. The organizational arrangement between the headquarters and subsidiaries changed frequently during this process. Knowledge was mainly transferred from the Japanese head office to overseas subsidiaries in the beginning, but this gradually turned into a 'learn locally and act globally' strategy in which local employees create local knowledge and competitive advantages (Ichijo & Kohlbacher, 2007).

The exact roles and boundaries between members in the Toyota production system are subject to the ongoing debate, which is precisely linked to the responsibilities that make continuous improvements possible. Learning by monitoring enables constructive discussion among members by uncovering differences in viewpoints through the exchange of information (Sabel, 1994). Building dynamic capabilities through voluntary participation from the shop-floor follows a sequential and cumulative cycle model of 'recombination → evaluation → knowledge sharing and problem solving → routinization' (Zolo & Winter, 2002: 343). Dorf and Sabel (1998) argue that the Toyota production system based on learning by monitoring and continuous improvement can be regarded as an example of democratic experimentalism, although it is true that the Toyota Production System is to some extent embedded in overseas transplants as a 'learning bureaucracy' or 'democratic Taylorism' (Adler & Cole, 1993).

7.3 Two Sides of Hyundai Motor's Production System and Future Prospects

7.3.1 Two Sides of Hyundai Motor's Production System

In this book, we have tried to look at both sides of Hyundai Motor's story. This is because Hyundai Motor's remarkable success has also come at a cost. Negative aspects such as relatively low wages for non-regular workers and lower unit prices for small and medium-sized subcontractors lurk in the shadows behind the company's growth. In other words, Hyundai Motor's success has involved transferring the burden of uneven development to the business ecosystem in Korea.

We have conceptualized the two sides of Hyundai Motor's production system by using the term 'authoritarian experimentalism.' The unique Korean business structure known as the *chaebol* is the reason why these two seemingly contradictory elements can coexist. The *chaebols* have been a major driving force behind Korea's economic catch-up. The Hyundai Group received preferential support from the government during the process of industrialization. Hyundai Motor has grown into a huge business group while receiving a number of financial and tax benefits in an environment that restricts competition in the domestic market. The government introduced the 'designated vertical integration' system in the late 1970s, making it mandatory for large companies to form subcontracting relationships with SMEs. This laid the groundwork for a system of extended quasi-vertical integration in which Hyundai Motor has near complete control over parts makers. The state also supported labor policies including suppressing wages and repressive labor controls.

With the full support of Korea's developmental state, Hyundai Motor was able to mobilize internal and external resources to build organizational capabilities and grow through intensive problem-solving. The vertical decision-making structure that connects the head of the *chaebol* group to non-affiliated parts makers made it possible to mobilize resources and carry out quick and bold investment decisions that entailed substantial risk. This helped Hyundai Motor to gain control of the market and grow rapidly through preemptive investment in a changing environment. However, it is difficult to monitor the decision-making process from the outside due to the vertical hierarchy of *chaebol* groups. Unilateral decision-making without close external monitoring leads to the risk of overinvestment, and this burden is passed on to group affiliates and non-affiliated parts makers, workers employed by these companies, and even the Korean economy as a whole.

Hyundai Motor's excessive investment in China is a typical example of this. As the global competition to secure the vast Chinese market ramped up, Hyundai Motor decided to rapidly expand its local production capacity in China through swift and bold decision-making. However, as local Chinese companies started catching up and Hyundai Motor was slow to respond by releasing appropriate models, the company's local production in China had sharply declined by the end of the 2010s (Interviews with a Hyundai Motor executive in 2018). In the end, the Chinese venture was a substantial financial misstep, resulting in severe consequences for parts makers that entered China together in circumstances where they were fully dependent on Hyundai Motor. This demonstrates the danger of vertical decision-making by *chaebol* groups.

Hyundai Motor's skill-saving work organization also creates job instability and segmentation in the labor market by largely leaving worker skills out of the picture. Hyundai Motor solves most of the problems that could occur in mass production of new vehicles through engineer-led problem-solving in the pilot production stage (see Chapter 3). During the mass production stage, general-purpose automation equipment is employed as much as possible to utilize the flexible potential of machines, while the work of shop-floor workers is standardized and simplified as much as possible (see Chapter 4). In this system, labor has been replaced by machines or became the target of numerical flexibility. It has become technically possible to replace shop-floor regular workers with non-regular workers under conditions that do not require a high degree of skill, and the proportion of non-regular workers has actually increased over time. The development of modular production has also led to a significant part of the internal production process being outsourced to modular parts makers, where mostly non-regular workers are responsible for the sub-assembly of modular parts.

The substitutability due to the absence of skilled workers has also led to an increased sense of job insecurity among regular workers, who are motivated to focus on protecting their jobs and increasing their short-term income through labor unions. Management has expanded the use of non-regular workers and outsourcing to cut costs and utilize numerical flexibility while maintaining the employment and wages of regular employees. In addition, the company has kept the wage of non-regular workers and the unit price of parts low on the pretext of labor union demands for employment security and higher wages. As a result, segmentation has occurred between regular workers and non-regular workers in Hyundai Motor's internal labor market, with the gap in wages and working conditions between widening over time. The system of extended quasi-vertical integration has intensified labor market segmentation between the automaker and parts makers, first-tier and second-tier component suppliers, and affiliates and non-affiliates, which is one of the negative aspects of Hyundai Motor's production system.

Hyundai Motor's extended quasi-vertical integration model effectively integrates not only affiliates but also non-affiliates, making it difficult for parts makers to act independently due to their almost total reliance on Hyundai Motor. This has allowed Hyundai to obtain absolute commitment from parts makers even while making unreasonable demands, which has served as an important driving force behind the company's growth (see Chapter 5).

Parts makers have been forced to provide parts at below market price in response to Hyundai Motor's demand. The wage rates of parts makers were differentiated by a hierarchy of subcontracting layers. The profits of parts makers have been limited to a certain level, with any additional profits boiling down to Hyundai Motor. Periodical unit price cuts place a heavy burden on parts makers, who are constrained in their ability to pursue independent growth by diversifying their customer channels. It is not easy to invest in new technology development or capability building when profitability is kept at such a low level. In addition, the hierarchy and gaps between companies with different subcontracting status have deepened as the parts supply structure has become more stratified due to modular production. In particular, the workshops of sub-vendor parts makers at the bottom of the subcontracting structure have effectively become sweatshops with poor working conditions and unstable jobs that pay low wages.

7.3.2 Future Prospects

Will Hyundai Motor's growth continue to be sustainable in the future? In this section, we prospect the future of Hyundai Motor as the company faces a fundamental transformation in the automobile industry. The automobile industry is currently in the midst of a tectonic shift comparable to the advent of the internal combustion engine car in the early twentieth century. The key features of this shift are the emergence of future vehicles (electric vehicles and hydrogen fuel cell vehicles) due to changes in power generation and transmission devices, advancement in software based autonomous vehicles, and mobility services and vehicle sharing (Meritz Securities, 2021).

The cars of the future are not simply shifting from internal combustion engines to batteries or fuel cells, but from hardware to software. Cars are becoming software-controlled vehicles, or software defined vehicles (SDVs). Of course, the tradition of manufacturing hardware products in large quantities is not going away, but the weight of production is declining in the automobile industry value chain, and the connection between mobility services and software that controls hardware and vehicles is emerging as a core component of competitiveness.

The transformation in the automobile industry has accelerated since the Volkswagen emissions scandal of 2015, with a particular focus on electric vehicles. As the technological innovation that purported to make the engines of internal combustion vehicles more eco-friendly has been proven false, the pave has picked up in the commercialization of future vehicles. Looking at the global outlook for electric vehicles, it is predicted that sales will exceed 28 million units in 2030 and that 57% of new cars sold will be electric vehicles by 2040, including plug-ins (BloombergNEF, 2022).

Tesla is the leader in driving this transformation of the automotive industry. The company uses an 'integral modular architecture' to control its electric vehicles, which are powered by cost-effective batteries, and makes frequent over the air (OTA) updates. In addition, a network of infrastructure, 6000 satellites, manufacturers and all vehicles in operation are connected with one another and able to communicate in real time. Tesla plans to commercialize a 'robotaxi' equipped with Level 4 autonomous driving (Meritz Securities, 2021).

Traditional automakers such as Hyundai Motor will be able to survive if they can appropriately respond to this trend. Otherwise they will face elimination from the market, or be relegated to mere subcontractors that supply hardware products. The driving force behind Hyundai Motor's production system is the improvisational ability of engineers. When top management set a goal of catching up with more advanced companies, the engineers have demonstrated their ability to achieve that goal. Will this production system allow Hyundai Motor to succeed in achieving a competitive advantage in the changing landscape of the automobile industry?

So far, Hyundai Motor appears to be relatively successful in 'catching up' to the industrial shift toward future cars. Although building software-centered organizational capabilities is the biggest challenge, Hyundai Motor has set the new goal of catching up with Tesla and striving to achieve compressed growth by hiring new staff and retraining current employees. The Hyundai Motor Group currently holds a 9% share in the US electric vehicle market, which ranks second after Tesla (S&P Global Mobility, 2022). Hyundai Motor's rapid progress in electric vehicles is helped by the fact that Tesla has made it easier for latecomers to catch up by publicly releasing all electric vehicle patents for free.

By 2030, Hyundai Motor aims to occupy 5.1% of the global electric vehicle market by electrifying all six models of the Genesis brand as well as another eleven models including six SUVs and three passenger cars for the remaining Hyundai-Kia Motors brands. Hyundai Motor has developed the electric global modular platform (E-GMP), the company's first platform dedicated to electric vehicles, and plans to complete the lineup with an integrated modular architecture in 2025 to improve performance and profitability. (Interviews with a Hyundai Motor executive in 2022).

We have attempted to come up with two possible transition models for the future of the automobile industry (see Table 7.2). Since this is a sort of ideal type, the model presented could not exactly correspond to the actual automaker. In the first model, corporate governance is marked by owner leadership. The CEO is not only an engineer himself, but also exercises strong management rights as the majority owner of the company. In human resource management, high-tech firms and startups actively scout and recruit talent to help develop future car technology. Labor–management relations remain confrontational or the company pursues a policy of union-free management, and unnecessary employees are boldly laid off. In the inter-firm relationships that receive materials and parts, market transactions at arm's length are promoted to reduce costs, and the production system of the headquarters is applied to overseas production facilities. This growth model could be effective in terms of the competitiveness of individual automakers, but it excludes stakeholders on the business ecosystem level, resulting in a widening gap.

The second model of corporate governance features professional managerial leadership. Top management is led by a professional manager with a proven internal track record ability or who has been recruited from outside. In human resource management, training new staff, external recruitment, and employment succession are all treated as important. Accordingly, the system provides opportunities for reassignment and reemployment in the field of future cars through retraining for existing employees. Both engineers and production workers have the opportunity to make use of their experience in smart factories. This is because even if automation and digitization advance, the skill of production technicians in charge of preparing for the mass production of new products and maintaining automated facilities becomes more important. For the sourcing of materials and parts, horizontal cooperative relationships are developed

Table 7.2 Two models for future vehicle transformation in the automotive industry

	Model 1: Tesla model	*Model 2: Volkswagen model*
Corporate governance	Owner's leadership	Professional managerial leadership
Human resource management	Software-centered development staff recruited, existing staff reorganized or laid off	Retraining and utilization of existing staff, links between engineers and production technicians
Work organization and labor–management relations	Unions excluded, smart factories, confrontational labor–management relations	Union participation, smart factories, labor–management partnership
Inter-firm relations	Arm's length transactions	Horizontal cooperation
Overseas production	Production system applied as is	Adaptation to local institutional conditions
Business ecosystem	Participation excluded, gap likely to widen	Participation encouraged, gap reduced

Source Author's own creation

and maintained with competent technology companies. When it comes to overseas production, even if the head office's production system is transferred to offshore facilities, the level of adaptation is increased by considering local institutional conditions if possible. This growth model involves reducing the gap by involving stakeholders on the ecosystem level while also improving the competitiveness of individual automakers.

It appears that both of these routes are possible in the transition to cars of the future. Although the first model is most in line with Tesla located at the forefront of electric vehicles, the second model is closest to Volkswagen, the automaker that has so far best adapted to industry transformation. The first model seems more likely at present, but it is highly unstable due to the heavy reliance on market functions, the requisite sacrifice of the business ecosystem to which the company belongs, and the CEO's discretionary decision-making. If the adaptation to future car technology proceeds smoothly, the second model appears more likely to generate a competitive advantage. In reality, there is also the possibility that a hybrid model with aspects of both models will become dominant. The future of the automotive industry is a story that is yet to be written.

Which of the two models will Hyundai Motor choose? Given the history of Hyundai Motor's production system, the first model appears to be the more likely candidate. However, this would only serve to widen the gap between individual companies and the rest of the Korean economy.

From this point of view, we believe Hyundai Motor must implement several changes in order to adapt to the industry transformation. First, in order to respond to changes in the business environment, Hyundai Motor should promote democratic participation through closer communication with employees, while at the same time maintaining the reasonable but strong *chaebol* head's leadership. Second, if new software, AI, and semiconductor experts are hired to promote the transition to car of the future, it is important that they are encouraged to draw on the wealth of experience among the company's engineers who majored in mechanical engineering. Third, Hyundai Motor should set up educational systems to improve the skills of the production technicians and skilled workers essential for running smart factories. To this end, efforts to improve labor–management relations should be carried out in parallel. Fourth, Hyundai Motor should attempt to transform inter-firm relationships into a more horizontal structure. This means enhancing the technological capabilities of parts makers through diversification of customers and better bargaining conditions. Finally, management of overseas plants needs to substantially increase the proportion of localization in product development and human resource management.

We believe Hyundai Motor will have to boldly accept these new challenges in order to keep up with the rapid pace of transformation in the automobile industry and emerge victorious. This means creating a new hybrid path of tradition and innovation based on the company's evolution to date. If the company is able to do so, Hyundai Motor's production system has the potential to develop into another 'best practice' for the industry.

References

Abo, T. (1994). *Hybrid Factory: The Japanese Production System in the United States*. Oxford University Press.

Adler, P. S., & Cole, R. E. (1993). Designed for Learning: A Tale of Two Auto Plants. *Sloan Management Review, 34*(3), 85–94.

Aghion, P., Guriev, S., & Jo, K. (2021). *Chaebols* and Firm Dynamics in Korea. *Economic Policy, 36*(108), 593–626.

Amsden, A. (1989). *Asia's Next Giant: South Korea and Late Industrialization*. Oxford University Press.

Aoki, M. (1988). *Information, Incentives and Bargaining in the Japanese Economy: A Micro-theory of the Japanese Economy*. Cambridge University Press.

Aoki, M. (1990). Toward an Economic Model of the Japanese Firm. *Journal of Economic Literature, 28*, 1–27.

Aoki, M. (2010). *Corporations in Evolving Diversity: Cognition, Governance, and Institutions*. Oxford University Press.

BloombergNEF. (2022). *Electric Vehicle Outlook 2022*. https://about.bnef.com/electric-vehicle-outlook/

Boyer, R., & Freyssenet, M. (2002). *The Productive Models: The Conditions of Profitability*. Palgrave.

Chandler, A. D. (2001). *Inventing the Electronic Century: The Epic Story of the Consumer Electronics and Computer Industries*. Free Press.

Dorf, M. C., & Sabel, C. F. (1998). A Constitution of Democratic Experimentalism. *Columbia Law Review, 98*(2), 267–473.

Fujimoto, T. (1999). *The Evolution of a Manufacturing System at Toyota*. Oxford University Press.

Fujimoto, T. (2006). Architecture-Based Comparative Advantage in Japan and Asia. In K. Ohno & T. Fujimoto (Eds.), *Industrialization of Developing Countries: Analysis by Japanese Economics* (pp. 1–10). National Graduate Institute of Policy Studies.

Gerschenkron, A. (1962). *Economic Backwardness in Historical Perspective*. Harvard University Press.

Gong, Y, T., Baker, T., & Miner, A. S. (2006). Capabilities and Routines in New Organizations: Evidence from the Field. (Working Paper School of Business, University of Wisconsin—Madison).

Gunasekaran, A. (1998). Agile Manufacturing: Enablers and an Implementation Framework. *International Journal of Production Research, 36*(5), 1223–1247.

Hattori, T. (1986). Technology Transfer and Management Systems. *The Developing Economies, 24*(4), 314–325.

Hattori, T. (2005). *Economic Sociology of Development: Korea's Economic Development and Social Changes*. Bunshindo Publishing. (In Japanese).

Herrigel, G. (2020). Experimentalist Systems in Manufacturing Multinationals: Recursivity and Continuous Learning through Destabilization. In J. Glückler, G. Herrigel, & M. Handke (Eds.), *Knowledge for Governance: Knowledge and Space* (Vol. 15, pp. 415–439). Springer.

Herrigel, G., & Sabel, C. F. (1999). Craft Production in Crisis: Industrial Restructuring in Germany during the 1990s. In Pepper D. Culpepper & David Finegold (Eds.), *The German Skills Machine: Sustaining Comparative Advantage in a Global Economy* (pp. 77–114). Berghahn Book.

Ichijo, K., & Kohlbacher, F. (2007). The Toyota Way of Global Knowledge Creation the 'Learn Local, Act Global' Strategy. *International Journal of Automotive Technology and Management, 7*(2–3), 116–134.

Janoski, T., & Lepadatu, D. (2021). *The Cambridge International Handbook of Lean Production: Diverging Theories and New Industries around the World*. Cambridge University Press.

Jo, H. J., & You, J.-S. (2011). Transferring Production Systems: An Institutionalist Account of Hyundai Motor Company in the United States. *Journal of East Asian Studies, 11*(1), 41–73.

Karabarbounis, L., & Neiman, B. (2013). The Global Decline of the Labor Share (NBER Working Paper, 19136).

Kim, L. (1997). *Imitation to Innovation: The Dynamics of Korea's Technological Learning*. Harvard Business School Press.

Lee, K. (2013). *Schumpeterian Analysis of Economic Catch-up: Knowledge, Path-Creation, and the Middle-Income Trap*. Cambridge University Press.

Lee, K., & Lim, C. (2001). Technological Regimes, Catching-up and Leapfrogging: The Findings from Korean Industries. *Research Policy, 30*(3), 459–483.

MacDuffie, J. P. (2013). Modularity-as-Property, Modularization-as-Process, and Modularity-as-Frame: Lessons from Product Architecture Initiatives in the Global Automotive Industry. *Global Strategy Journal, 3*, 8–40.

Meritz Securities. (2021). *AI War: Tesla vs Non-Tesla*. Meritz Securities.

Mosca, M. (2018). *Monopoly Power and Competition: The Italian Marginalist Perspective*. Edward Elgar.

Nagel, R. N. (1991). *21st Century Manufacturing Enterprise Strategy: An Industry-led View*. Lehigh University.

Naylor, J. B., Naim, M. M., & Berry, D. (1999). Leagility: Integrating the Lean and Agile Manufacturing Paradigms in the Total Supply Chain. *International Journal of Production Economics, 62*(1–2), 107–118.

Oh, J. (2022). *Research on the History of Technological Development of Hyundai Motor Company*. Meiji University. (In Korean).

S&P Global Mobility. (2022). New EV entries nibbling away at Tesla EV share, according to S&P Global Mobility, Press Releases (https://press.spglobal.com/2022-11-29-New-EV-entries-nibbling-away-at-Tesla-EV-share,-according-to-S-P-Global-Mobility).

Sabel, C. F. (1994). Learning by Monitoring: The Institutions of Economic Development. In N. Smelser & R. Swedberg (Eds.), *Handbook of Economic Sociology* (pp. 137–165). Princeton University Press and Russell Sage Foundation.

Shintaku, J., Ogawa, K., & Yoshimoto, T. 2006. Architecture-based Approaches to International Standardization and Evolution of Business Models [MMRC Discussion Paper 96, University of Tokyo].

Sturgeon, T. J. (2002). Modular Production Networks: A New American Model of Industrial Organization. *Industrial and Corporate Change, 11*(3), 451–496.

Teece, D. J. (2017). Towards a Capability Theory of (Innovating) Firms: Implications for Management and Policy. *Cambridge Journal of Economics, 41*(3), 693–720.

Veblen, T. (1915). *Imperial Germany and the Industrial Revolution*. Macmillan.

Vernon, R. (1966). International Investment and International Trade in the Product Cycle. *The Quarterly Journal of Economics, 80*(2), 190–207.

Whittaker, D. H., Sturgeon, T. J., Okita, T., & Zhu, T. (2020). *Compressed Development: Time and Timing in Economic and Social Development*. Oxford University Press.

Womack, J. P., Jones, D. T., & Roos, D. (1990). *The Machine that Changed the World*. HarperPerennial.

Zollo, M., & Winter, S. G. (2002). Deliberate Learning and the Evolution of Dynamic Capabilities. *Organization Science, 13*, 339–351.

Index

A

Abo, Tetsuo, 35, 198, 199, 219, 238

Acquisition, 66, 73, 152, 178, 240, 241

Adaptation, adapt, 30, 32, 35, 81, 148, 189, 199–204, 216, 249, 250

Advantageous opportunity, 47, 48

Affiliate, 30, 31, 34, 36, 45, 48, 79, 80, 83–85, 87, 98, 108, 110, 111, 120, 149, 153, 154, 171, 172, 175, 177, 179, 182, 183, 185, 186, 192, 193, 200, 205–207, 229, 230, 237, 239, 241, 244–246

After-sales, 178, 179, 187

Agile, 48, 52, 231–233

- agile production system, 46–49, 122, 142, 144, 147, 171, 199, 201, 202, 204, 216, 219, 225, 233, 234, 238, 239

Amsden, Alice, 5–8, 12, 28, 29, 32, 33, 99, 119, 121, 139, 227–229

Aoki, 7, 8, 23, 27, 41, 42, 77, 242

Application, apply, 15, 35, 70, 96, 106, 122, 164, 199, 204

Appropriation of profit, 83

Approved drawings, 181

Asan, Asan plant, 68, 150, 151, 178, 201

Assembly, 9, 38, 66, 76, 104, 130, 132, 133, 135, 149, 150, 153, 154, 156, 157, 159, 164, 165, 173, 185, 186, 190, 201, 208, 209, 217, 231, 234, 237

- Assembly Line Control (ALC), 70, 150, 157, 165
- assembly-type industrialization, 9, 232

Authoritarianism, 8

Authoritarian state, 31

Automation, 9, 32, 38, 45, 49, 102–104, 106–108, 141, 149–151, 154, 156, 163, 167, 175, 201, 203, 211, 235–237, 239, 240, 245, 248

- labor-replacing automation, 150

H. J. Jo et al., *Agile Against Lean*,
https://doi.org/10.1007/978-981-99-2042-6

Automobile industry, 3, 4, 35, 36, 39, 46, 49, 73, 75, 76, 92, 96, 98, 103, 108, 111, 119, 137, 147–149, 153, 166, 173, 176, 177, 185, 189, 192, 194, 225, 228, 246–248, 250
Autonomous driving, 247
 autonomous (driving) vehicle, 194, 225, 246
Autonomy, 7, 27, 28, 52, 66, 74, 100, 102, 126, 147, 210, 220, 225, 226, 229, 231, 234, 241, 242

B

Bargaining power, 69, 84, 107, 109, 181, 230
Benchmark, benchmarking, 26, 43, 45, 49, 68, 70, 71, 76, 230–232, 239
Best practices, 44, 212, 230, 231, 250
Big firm, 15, 20–24, 26, 27, 44, 45, 74, 99, 231
Black box, 225, 227
Body, 66, 68, 74, 133, 165, 201, 208, 211, 237
Boyer, R., 35, 37, 40, 42, 86, 230, 233
Bureaucratic control, 187, 189

C

Candidate, 188, 250
Capability, 8, 21, 22, 26, 29, 43, 76, 120–122, 124, 126, 129, 130, 133, 139, 144, 179, 185, 207, 231, 241, 246
 capability building, 5, 6, 20, 21, 45, 49, 74, 127, 227, 240, 241
 organizational capabilities, 20–23, 26, 45, 48, 99–101, 111, 120, 126, 136, 140, 143, 166, 208, 210, 212, 217–220, 234, 235, 238, 241, 244, 247
Captive, 175, 176, 188–190, 239, 240
Carrots and stick, 31, 240
Catch(ing)-up, catch up, 4–6, 8, 10, 12, 32, 43, 45, 46, 49, 70, 71, 73, 76, 77, 102, 140, 153, 193, 219, 226, 227, 231, 238, 244, 247, 248
 catch-up 1.0, 65, 227
 catch-up 2.0, 65, 227, 228, 231, 232
Chaebol, 4, 7, 8, 10–12, 20, 28–32, 34, 42, 43, 46, 48, 52, 66, 74, 76–84, 86, 95, 101, 102, 108–111, 120, 128, 172, 176, 177, 181, 186, 200, 225–230, 232, 234, 235, 240–242, 244, 245, 250
Champion, 188
Chandler, A.D., 16–20, 45, 186, 229
 Chandlerian, 16, 18, 19, 23, 28, 44, 45, 48, 50, 52, 70, 74, 84, 226, 228, 229, 231
 post-Chandlerian, 27, 28, 44, 45, 48, 50, 74, 226, 228, 231, 232
 pre-Chandlerian, 28, 48, 50, 52, 74, 84, 226, 229, 231
Chassis (module), 154, 155, 177–179
Cheap and high-quality car, 74
Chongsu, 29, 34, 78–83, 123, 200, 217, 229, 238, 240, 241
Cockpit (module), 154, 155, 178
Codification, 28, 76
Collaboration, 15, 26, 48, 143, 194, 228
Competitive advantage, 10, 17, 18, 21, 24–26, 28, 46, 49, 69, 73, 139, 166, 167, 230–232, 237, 243, 247, 249
Compressed growth, 8, 225, 247

Concept, conception, 5, 8, 13, 28, 29, 32, 37, 40, 42, 43, 49–51, 86, 100, 106, 119, 127, 148, 163, 193, 199, 201, 203, 204, 216–220, 230, 236, 238–240
Conglomerate, 19, 86
Control, 3, 7, 14, 18–20, 25, 28, 30–32, 34, 35, 37, 38, 40, 44, 65, 76–80, 83, 100, 102, 108–110, 140, 161, 181, 185, 188, 189, 192, 229, 231, 237, 239, 241, 242, 244, 247
Cooperation, 12, 13, 15, 23, 25, 27, 28, 44, 47, 66, 78, 79, 101, 139, 140, 179, 231, 234, 237, 249
Coordinated market economy, 41
Coordination, 6, 7, 13, 15, 20, 27–30, 44, 50, 76, 78, 81, 110, 131, 229, 231
Corporation, 16, 19, 41, 50, 66, 78, 228, 229, 231
  large corporation, 9, 12, 16, 17, 19–21, 24–27, 32, 44, 51
Cost-linked pricing, 110
Cost reduction, 41, 67, 86, 98, 152, 177, 185, 193, 240, 243
Creative destruction, 16
Cross Functional Team (CFT), 128, 131–138, 144, 235, 241
Cross-shareholding, 30, 80, 83, 84, 192, 200
Cumulative, 6, 20, 21, 40, 52, 95, 126, 136, 138, 231, 232, 241, 243

D

Daewoo, 189
Daily (plant or factory) operation, 199, 204, 211, 212, 214–216
Decision-making, 27, 32, 34, 46, 48, 51, 111, 123, 125, 136, 143, 200, 217, 220, 229, 244, 245, 249
Design, 9, 11, 12, 33, 36, 39, 49, 66–68, 76, 95–97, 99, 101, 102, 120, 121, 125, 127–138, 140–142, 144, 148, 153, 161, 165, 171, 179, 185, 188, 201–203, 205, 208–210, 212, 213, 218, 226, 233, 234, 237
Developed country, 3, 4, 6, 32, 75, 112, 173, 226, 227
Developing country, 3, 4, 67
Developmental state, 6–8, 162, 173, 190, 244
Dicken, Peter, 172, 192
Direct control, 34, 182
Direct investment, 73, 75, 111, 200, 204
Discretion, 47–49, 70, 129, 133, 144, 233
Distrust, 38, 71, 73, 122, 157, 162, 167, 236
Diversification, 28–31, 81, 89, 120, 121, 250
Diversity, 28, 35, 40, 42, 47, 86, 92, 107, 230, 233, 240
Division of labor, 9, 17, 20, 37, 38, 41, 47, 84, 142, 147, 217, 226, 227, 236, 239
Door shield (module), 154, 155
Downsizing, 185

E

Economy of scale (and scope), 5, 9, 11, 12, 14, 18–20, 24, 28, 30, 38, 45, 47, 69, 73, 74, 80, 89, 103, 110, 111, 122, 124, 186, 197, 206, 207, 236, 241
Electric global modular platform (E-GMP), 248
Electric vehicle (EV), 49, 89, 92, 103, 194, 246–249

Employment relationship, 41, 42
employment adjustment, 106
employment system, 36, 104, 237
Engineer, 4, 10, 32, 33, 37–39, 45, 48, 49, 51, 52, 65–67, 70, 74, 76, 95–102, 106, 107, 111, 119–126, 128–131, 133–144, 147, 149, 156, 158, 161, 162, 166, 167, 199, 208, 211, 212, 225, 226, 229, 231, 232, 234–236, 238, 241, 247–250
engineer-led, 49, 99, 128, 130, 154, 166, 171, 232, 238–240, 245
Enterprise Resource Planning (ERP), 157, 158
Entrepreneurial, entrepreneurship, 13, 15, 17–19, 22, 28–31, 46, 48, 49, 70, 77, 83, 227–229, 238, 241, 242
Equity, 73, 79, 87, 172, 176, 186, 189, 191, 192
Evaluation, 50, 187, 188, 213
Evaluation Group, 136
Evolution, Evolutionary, 18, 22, 40, 44, 45, 52, 70, 75, 228, 231, 250
Excel, 67, 68, 70, 149
Execution, 22, 28, 32, 38, 49–51, 106, 120, 121, 126, 127, 143, 163, 174, 230–232, 236, 239, 240, 242
Expand, expansion, 20, 75, 200, 244
Experimentalism, 27, 51, 52, 101, 143, 229, 231, 234, 238
authoritarian experimentalism, 46, 50, 52, 123, 125, 226, 229, 234, 240, 242, 244
democratic experimentalism, 50–52, 240, 243
Exploitation, 37, 191, 193, 232, 236
Export, 7, 29, 67, 68, 71, 74, 102, 139, 149, 161, 173, 178, 187, 191, 197, 201, 210, 219, 243

F

Fast follower, 87
Finance, financial, 7, 10, 12–14, 16, 19, 20, 25, 30, 31, 40, 41, 43, 44, 65, 71, 77, 79, 84, 86, 97, 102, 107, 122, 152, 172, 173, 175–178, 191, 204, 228, 229, 234, 236, 244, 245
financialization, 42
Financial crisis
Asian financial crisis, 4, 11, 28, 30–33, 65, 73, 76, 86, 87, 89, 96, 97, 99, 101, 103, 104, 106, 107, 225, 227, 228, 230, 234, 236
Global financial crisis, 4, 47, 87
Firm-specific skill, 25, 32, 68, 69, 106
First-tier, 76, 153, 173, 174, 176, 182, 183, 187, 205, 206, 215, 245
5(Five) star certification, 187, 188
Flexibility, 40, 47, 70, 84, 86, 87, 98, 100, 107, 111, 112, 148, 152–154, 157, 163, 165, 166, 177, 185, 191, 230, 233, 236, 239, 241
Flexible
flexible automation, 69, 106, 107, 148, 150, 152, 157, 165, 166, 225, 240
flexible production, 102, 107, 147, 148, 150, 151, 157, 162, 167, 199, 216, 219
flexible specialization, 15, 147
flexible standardization, 148
Flexible Body Line (FBL), 70, 151
Follow(ing) sourcing, 199, 205–207, 215, 216, 218, 220, 238, 241

Fool proof, 159
Ford, 38, 66, 74, 149, 200, 230, 232
Fordism, 15, 37–40, 104, 148, 166, 230, 235, 236, 239
Foreman, 158
Free market economy, 41
Freyssenet, Michel, 35, 37, 40, 42, 86, 230, 233
Front end module (FEM), 154, 155
Fujimoto, Takahiro, 10, 25, 27, 39, 40, 45, 100, 120, 129, 186, 200, 217, 229, 231
Full-employment protection agreement, 106, 236

G
General-purpose equipment, 102, 150
  general-purpose facility, 130
Gerschenkron, A., 5, 6, 8, 11, 43, 49, 73, 226, 227
Global Human Resource Standard (GHRS), 204, 205, 214
Globalization, 42, 43, 75, 77, 111, 217, 227, 237, 241
Global Top-Five (GT-5), 73
Global value chain (GVC), 4, 45, 77, 188, 199, 207
Governance, 4, 8, 11, 19, 23, 27, 34, 37, 41–43, 46, 76–79, 81, 83, 84, 86, 99, 187, 188, 200, 219, 225, 229, 238, 240, 242, 248, 249
Government, 6, 7, 9–13, 16, 23, 29, 30, 32, 46, 66, 69, 73, 77, 149, 163, 173, 193, 202, 226, 244
Great Labor Struggle of 1987, 31
Green field, 199
Groupe d' Etudes et de Recherches Permanent sur l' Industrie et les Salariés de l'Automobile (GERPISA), 35, 37, 40
Guest engineer, 134

H
Hattori, Tamio, 9, 33, 100, 190, 232
Headlining, 154, 155, 175
Headquarter, 31, 97, 123, 204, 207, 208, 211, 212, 217–219, 238, 243, 248
  Headquarter-subsidiary relationship, 199
Hierarchy, 18–20, 26, 27, 47, 50, 99, 101, 140, 141, 165, 172, 181–183, 191, 234, 244, 246
Honda, 86, 218
Hostile coexistence, 107
Hostile labor (management) relations, 37, 70, 107, 162, 167, 240
Human resource (HR), 16, 120, 214, 236
  human resource management (HRM), 38, 199, 200, 203, 204, 213, 214, 219, 220, 248–250
Hyundai
  Hyundai Group, 73, 121, 126, 149, 177, 244
  Hyundai Mobis, 36, 79, 80, 84, 87, 98, 153, 154, 156, 164, 171, 172, 177–186, 189, 193, 200, 206, 237, 238
  Hyundai Motor, 3–5, 36, 37, 42, 43, 45, 46, 48–52, 65–71, 73–77, 79, 80, 82, 84, 86, 87, 89, 92, 95–112, 119–127, 129, 130, 132, 134–144, 147–154, 156–167, 171–179, 181–194, 197–208, 210, 211, 214–219, 225–228, 230–248, 250
  Hyundai Motor Group, 73, 79–81, 83, 84, 89, 98, 123, 131, 152–154, 171, 177, 178, 183, 186, 189, 197, 198, 200, 201,

204–209, 212, 213, 217–220, 237, 239, 247
Hyundai Precision Industry, 153, 179, 184
Hyundai Cooperative, 174–176
Hyundai Motor's production system, 66, 111, 243
Hyundai Quality System (HQS), 187

I

Improvement, 21, 25–27, 32, 36, 39, 40, 45, 48, 51, 68, 69, 74, 75, 92, 95, 106, 122, 132, 134, 137–139, 141, 142, 147, 162, 185, 212, 220, 231–233, 243
Improvisation, improvisational, improvise, 22, 48, 52, 95, 100, 101, 120, 126, 127, 129, 133, 135, 136, 138, 144, 147, 225, 231, 232, 234, 238, 241, 247
Indirect worker, 106
Industrial district, 13, 15, 16, 20, 24
Industrialization, 7, 9, 11–13, 16, 24, 119, 121, 123, 137, 139, 140, 190, 244
Industrial upgrade, 181
Information, 26, 27, 32, 33, 41, 47, 65, 69, 106, 120, 125, 138, 149, 158, 177, 179, 186, 236, 243
information (and communication) technology, 46, 154, 228, 236
information system, 149, 150, 157
Informatization, 49, 141, 148, 149, 154, 156, 157, 166, 167, 235, 236, 240
In-house modularization, 186
Initial Quality Study (IQS), 70, 71, 92, 95
Innovation, 3, 5, 6, 11–13, 15–17, 19–21, 23–27, 33, 38, 40, 41, 45, 51, 74–76, 80, 86, 96, 100, 110, 121, 123, 136, 138, 140, 147, 171, 228, 247, 250
Insecurity
employment insecurity, 122
job insecurity, 71, 163, 245
Institutional complementarity, 35, 77, 104
Institution, institutional, institutionalist, 6, 7, 11–14, 16, 18, 23, 30, 41–43, 48, 49, 68, 70, 77, 198, 214, 226–228, 238, 242, 249
Intensive problem solving (capability), 128, 135, 137
Interdependent relationship, 171
Inter-firm (business) relationship, 17, 27, 28, 84, 96, 104, 190, 192, 193, 237, 240, 242, 248–250
Internalization, 87, 181
Internal transaction, 186
Investment
aggressive investment, 107, 200, 201, 204
bold investment, 29, 46, 200, 240, 241, 244
high-risk investment, 30
independent investment, 200, 217
investment-based, 11, 28, 227
investment-led growth, 71
overinvestment, 31, 244

J

Japan, Japanese, 3, 5, 7, 9, 23–27, 31, 33–36, 43, 46, 66–68, 75, 76, 96, 119, 121, 135, 139–143, 153, 172, 175, 185, 186, 189, 191, 192, 199–201, 207, 217–220, 228, 242, 243
Japanese production system, 67, 73, 121, 122, 137, 139, 141, 144
Job rotation, 25, 51, 106, 142, 158, 160, 165, 166, 213

Jürgens, Ulrich, 35, 37–42, 76, 78
Just-in-sequence(JIS), 178, 235, 239–241
Just-in-time (JIT), 5, 27, 39, 40, 51, 69, 76, 104, 231, 235, 239, 240, 242, 243

K
Kaizen, 36, 162
Keeper, 106
Keiretsu, 23, 25, 34, 40, 186, 191, 192, 231, 240, 242
Kia (Motors), 73, 75, 79, 80, 84, 89, 95, 96, 99, 111, 178, 179, 181, 203, 206, 215
Kim, Linsu, 125
Knocked-down, knock down (KD), 173, 219
Knowle(d)ge, 5, 14, 15, 20–22, 26–28, 32, 33, 37, 41, 43, 80, 84, 119, 139, 152, 233, 238, 243
  codified knowledge, 226
  explicit knowledge, 26
  tacit knowledge, 21, 24, 26, 129, 152, 227
Korea, Korean, 4, 6–11, 20, 28, 30–34, 36, 43, 66–71, 73, 75, 77, 78, 80, 87, 92, 96–98, 100, 103, 106, 108, 111, 119, 120, 123, 139, 140, 149, 152, 162, 163, 167, 173, 176, 177, 186, 189–191, 198–203, 205, 206, 210–216, 218, 219, 228, 229, 232, 237–239, 241, 244, 250

L
Labor control, 31, 38, 69, 162, 163, 167, 244
Labor (-management) relations, 36, 71, 96, 106, 112, 122, 142, 148, 150, 157, 162–165, 167, 204, 213, 216, 235, 240, 248–250
  confrontational labor (-management) relations, 49, 69, 103, 109, 143, 144, 154, 156, 158, 162, 163, 213, 249
  hostile labor (-management) relations, 37, 70, 107, 162, 240
Labor market, 32, 68, 98, 106, 148, 165, 236, 245
  internal labor market, 69, 165, 167, 245
Labor process, 38, 40, 50, 167, 239
  work process, 37, 127, 148, 239
Latecomer, 4–7, 12, 24, 43, 45, 102, 121, 122, 125, 137, 154, 166, 219, 226–228, 231, 238, 248
Layoff, 71, 122, 236
Lazonick, William, 13, 15, 19, 20, 25, 27, 32, 44
Lean production (system), 3, 21, 25, 26, 35, 36, 39, 46–51, 68, 70, 71, 76, 147, 153, 162, 165, 185, 230, 231, 233, 238, 239
Leapfrog, 43, 52, 67, 226, 235
Learning, 4, 5, 9, 12, 13, 15, 20–22, 24, 25, 30, 32, 33, 39–41, 52, 66, 70, 76, 101, 125, 213, 227–229, 231, 232, 234, 235, 240, 241, 243
Learning bureaucracy, 39, 50, 229
Lee, Keun, 6, 10, 68, 227, 228
Linear, 6, 232
Localization, 173, 200, 204, 211, 212, 215–220, 250
Long-Term Automobile Industry Promotion Plan, 66, 149, 173
Low-priced, 68, 73
Low-quality, 70, 73

M

Mabuk-ri Research Center, 67
MacDuffie, John Paul, 171, 172, 232
Machine technology, 219
Maintenance, 70
Manufacturing design technology, 138, 140
Manufacturing Design (Technology) Center, 201
Markup, 86, 87, 89, 91, 92, 107
Marshall, Alfred, 12–14, 16, 17, 19
Marshallian, 12, 20, 21, 23, 24, 44, 228
Mass production, 5, 24, 38, 46, 48, 50, 67, 96, 101, 102, 120, 121, 124, 129–134, 139–142, 147, 161, 165, 167, 208–211, 219, 220, 230, 231, 234, 237, 239, 245, 248
mass production process, 147, 148
mass production system, 37–40, 45, 47, 67, 68, 71, 104, 139, 173, 230, 231
Material Requirement Plan (MRP), 102, 149, 150, 157
Mergers and acquisitions (M&As), 73, 74, 86, 112, 182
Middle-ranked maker's trap, 121
Minimal structure, 100
Mistrust, 69
Mitsubishi, 4, 66–68, 74, 121, 125, 137, 139, 140, 200, 230
Mobility service, 246, 247
Model factory, 199, 201–204, 216, 218–220, 238
Module
modular component, 161, 177, 181, 186
modularity, 76, 185
modularization, 12, 28, 34, 36, 45, 48–50, 65, 76, 77, 84, 107, 110, 134, 152–154, 161, 171, 182, 183, 185, 193, 208, 226, 228, 230–232, 237, 239–242
modularize, 156, 226, 232, 235
modular parts, modular parts supplier, 76, 84, 153, 154, 156, 164, 171, 176–179, 181, 182, 184–186, 189, 191–193, 237–239, 245
modular production, 36, 121, 148, 152–154, 156, 157, 164, 165, 172, 177, 178, 181, 185, 186, 192, 193, 245, 246
Monopoly, monopsony, 14, 29, 34, 46, 48, 73, 87, 99, 108, 110, 177, 178, 189, 230, 237, 241
Multi-functional, 25, 51
multi-skilled, 25, 40, 51, 142, 158, 159, 165, 166, 213, 220, 242
Multilayered, 183
Multiple orders, 34, 110, 237
Mutual obligation, 192

N

Namyang Research (and Development) Center, Namyang R&D Center, 77, 121, 123, 161, 179, 208, 210, 217, 234, 235, 238, 240
Neo-Schumpeterian, 18, 20, 22, 23, 44, 45, 70, 228
Network, 5, 12–14, 18, 23, 26–28, 35, 70, 71, 81, 84, 110, 111, 171, 172, 176, 183, 188, 189, 191–193, 242, 243, 247
network company, 228
network with network, 172
New car development, 101, 121, 125, 130, 133, 135–140, 142, 144, 161, 187, 199, 205, 207, 209, 210, 218, 232, 235, 238
New production concepts, 147

Non-affiliate, 172, 175, 176, 181–184, 189–193, 205, 206, 237, 244–246
Non-linear, 6, 126
Non-regular worker, 106–108, 156, 159, 164, 165, 167, 236, 244, 245
Non-sequential, 19, 232, 241
Norm, 188, 204
Numerical flexibility, 32, 156, 163, 164, 167, 245

O
Off-the-job training (Off-JT), 151, 166
Oligopoly, 3, 73, 74, 189
On-the-job training (OJT), 38, 151, 166, 212, 242
Oppressive control, 31, 38, 162
Organizational culture, 126, 133, 156, 203
Outsource, outsourcing, 36, 76, 84, 98, 154, 156, 164, 165, 177, 185, 186, 230, 237, 245
Overseas, 4, 46, 74, 75, 111, 121, 124, 143, 167, 175, 177, 190, 193, 197–202, 204, 205, 210, 212, 213, 217, 218, 220, 235, 237, 238, 241, 243
  overseas plant, 112, 122, 123, 150, 167, 197–205, 207–220, 234, 238, 243, 250
  overseas production, 75, 96, 111, 167, 176, 191, 197–202, 219, 238, 241, 248, 249
Owner, 20, 30, 217, 229, 248, 249
  ownership, 18, 19, 23, 28, 30, 48, 50, 52, 74, 79, 80, 82, 84, 172, 176, 189, 191, 192, 200, 226, 229, 231, 242

P
Parent company, 120, 173–176, 186, 200
Parkin, F., 181
Participation, 25, 36, 40, 51, 52, 68, 74, 77, 101, 104, 107, 122, 138, 144, 157, 162, 163, 167, 188, 203, 220, 231, 232, 234, 239, 240, 242, 243, 249, 250
Partner, 15, 34, 51, 110–112, 134, 188, 200, 232, 239, 241
Parts, 4, 9, 12, 24, 34, 36, 38, 40, 48, 66, 67, 69, 75, 76, 84, 87, 98, 104, 108–110, 120, 125, 131, 132, 134, 142, 149, 150, 153, 154, 156, 157, 159, 161, 164, 165, 173–179, 181–193, 197–199, 205–208, 210, 211, 213, 215, 216, 218, 219, 226, 230, 232, 235, 237–239, 241, 242, 244–246, 248, 250
Parts supplier, 49, 76, 102, 107, 132–135, 143, 144, 153, 156, 161, 171, 172, 175, 178, 182–185, 187–191, 193, 194, 206, 207, 210, 211, 215, 218, 220, 229, 231, 232, 234, 237, 239, 243
  parts supply network, 40, 182, 237
  parts supply structure, 171, 181, 182, 185, 189, 237, 239, 246
Paternalistic control, 175, 176
Path dependency, path dependence, 198, 199, 213, 219
Personnel, 19, 133, 135, 149, 179
  personnel exchange, 141, 142, 192
  personnel management, 214
  personnel system, 141, 214, 216
Pilot center, 48, 51, 52, 77, 102, 121–125, 127–137, 142–144, 147, 209, 210, 232, 234, 235, 238–241

pilot car, 123, 124, 129–131, 133–135, 209, 210
pilot production, 76, 123–125, 129, 130, 132–135, 138, 144, 161, 209, 210, 220, 245
Platform, 67, 89, 92, 208, 240, 248
platform integration, 92, 239, 240
platform sharing, 74, 153
Pony, 66, 149
Precision Industries, 153, 174, 177
Pre-Delivery Inspection (PDI), 162
Price cuts, 34, 109, 176, 232, 246
Proactive improvement, 215–216
Problem solve, problem solving, 22, 26, 52, 126, 135, 147, 161
Process technology, 11, 33, 51, 96, 121, 125, 129, 131, 137–142, 156, 234
Procure (ment), 24, 76, 173, 190, 215, 216, 219, 238
Product architecture, 125, 226
Product development, 22, 40, 47, 48, 75, 96, 98, 120, 125, 127–130, 134, 136, 138, 139, 153, 156, 175, 210, 217, 218, 220, 232, 240, 250
Production technology, 32, 68, 139–141, 148, 149, 153, 156, 165, 166, 201, 239
Productive model, 230, 233
Profit, 18, 29, 37, 40, 42, 67, 77, 78, 83, 84, 86, 89, 92, 103, 110, 136, 164, 179, 180, 183, 185, 205, 240, 243, 246
Project execution capability, 29
Purchasing, 131, 157, 175, 187

Q
QS9000, 187
Quality
Global Quality Management System, 74
quality assurance, 133, 175, 187, 191, 193
quality control (QC), 67, 106, 141, 161, 187, 188, 211, 235
quality evaluation system, 187–189
Quality (Management) Headquarters, 74, 123, 161, 187
quality management (system), 48, 74, 101, 111, 148, 161, 166, 167, 176, 234, 241
Quality control (QC), 67, 106, 139, 141, 161, 187, 188, 211, 235
Quasi-vertical integration, 34, 79–81, 84, 87, 108, 171, 172, 176, 191, 228, 232
extended quasi-vertical integration, 34, 83, 108, 110, 171, 172, 176, 177, 182, 183, 185, 188–194, 230–232, 237, 238, 241, 244–246
quasi-integrated inter-firm relationships, 171

R
Real-time planning, 126
Recombination, 100, 128
Relationship-specific, 81
Rental drawings, 181
Rent-seeking, 29, 46
Research and development (R&D), 5, 7, 11–13, 32, 33, 67, 74, 96, 121, 130, 179, 187, 209, 232
Restructuring, 30, 67, 73, 76, 77, 148, 176, 239
Return on equity (ROE), 70, 86, 87, 89
Revenue, 86, 181, 183
Reverse engineering, 6, 125, 139
Risk-taking, 229, 241
Robot, 103, 104, 149, 151, 165, 211
Roof (module), 133, 154, 211

roundabout cross-shareholding, 30
Roundabout shareholding, 80
Routine, 18, 20–22, 25–27, 40, 51, 70, 100, 101, 120, 126, 127, 136, 138, 144, 220, 231, 241

S
Schumpeter, J.A., 16, 17, 20, 45
  Neo-Schumpeterian, 18, 20, 22, 23, 44, 45, 228
  Schumpeterian, 12, 16, 24
Second-tier, 182, 183, 188, 206, 215, 245
Segmentation, 21, 98, 157, 165, 167, 245
  labor market segmentation, 162
  segmented labor, 102, 225
Selective affinity, 154, 203
Sequential, 6, 21, 43, 52, 73, 95, 101, 125, 126, 227, 235, 239, 241, 243
Shareholder, 19, 30, 31, 41, 42, 77–79, 82, 86, 87, 229
Shop-floor, 12, 21, 33, 36, 38, 45, 49, 51, 52, 69, 70, 100, 102, 106, 122, 139–143, 147, 157, 158, 161, 163, 166, 167, 214, 234, 236, 238, 239, 242, 243
  shop-floor worker, 20, 25, 27, 33, 37, 39, 40, 45, 49, 51, 68, 95, 100, 106, 121, 122, 124, 130, 131, 135, 137–142, 150, 152, 154, 156–159, 161, 162, 165–167, 213, 214, 225, 230, 232, 234, 236, 240, 245
Short-termism, 34, 47, 78, 87, 95, 107, 185, 229, 245
Simplification, 103, 235, 237
Simultaneous engineering, 26, 232
Single best way, 35, 37, 78
Skill, 20, 25, 27, 33, 36, 51, 71, 78, 106, 107, 119, 120, 126, 141, 144, 147, 148, 151, 159, 166, 203, 213, 232, 236, 245, 248
  skilled worker, 24, 25, 51, 100, 135, 141, 156, 161, 165, 167, 209, 211, 232, 245, 250
  skill formation, 7, 23, 45, 49, 76, 78, 100, 125, 142, 147, 150, 158, 159, 166, 213, 232
  skill qualification system, 141, 151, 152
Skill-saving work organization, 144, 147, 148, 152, 158, 162–167, 171, 199, 200, 202, 203, 213, 214, 216, 219, 245
Small and Medium Business Subcontracting Promotion Act, 173
Small and medium-sized enterprises (SMEs), 9, 12, 13, 16, 23, 44, 183, 184, 188, 190, 228, 244
Social closure, 181
Socialization of loss, 83
Software, 102, 125, 150, 157, 246, 247, 249, 250
  software defined vehicles (SDVs), 246
Spatial fix, 112, 237
Stakeholder, 31, 41, 42, 48, 51, 52, 74, 77, 78, 81, 101, 102, 229, 231, 234, 235, 248, 249
Standards, 3, 4, 6, 21, 26, 38, 39, 76, 97, 133, 136, 138, 188, 230, 236
  standardization, 11, 12, 26, 28, 38, 49, 132, 154, 208, 217, 226, 235, 239
Strategy, 9–11, 13, 20, 28, 30, 32, 37, 38, 40–43, 48, 66, 68, 69, 71, 84, 86, 87, 89, 92, 95, 96, 106, 112, 121, 123, 127, 139, 150, 152–154, 156, 164, 173, 177, 199–201, 217, 218, 220,

231–233, 237, 238, 240, 241, 243
Strike, 69, 107, 122, 213
Subcontract, 87
subcontracting relationship, 173, 191, 242, 244
subcontracting system, 173, 175, 190, 191, 193, 242
subcontractor, 25, 42, 76, 110, 165, 173, 230, 241, 244, 247
Subsidiary, 29, 79, 80, 84, 111, 128, 177, 204, 214, 220, 243
Supplier, 14, 28, 32, 34, 40, 41, 45, 76, 84, 108, 109, 133, 143, 154, 164, 171–177, 183–187, 190, 205, 206, 220, 225, 237–239, 242, 245
suppliers' council, 173, 174, 182
Supplier Quality Mark (SQM), 188
Supply chain, 171, 172, 176, 182, 185, 188, 191, 192, 215, 233
supply network, 181, 188, 194
Sweatshop, 246
Synchronize, synchronization, 51, 104, 233, 242
Synecthry, 107

T
Taylorism, 37–39, 106, 230, 239, 242
democratic Taylorism, 39, 50, 229
Technician, 33, 121, 139–141, 152, 248–250
Technocratic, 45, 48, 49
Technological capabilities, 6, 32, 66, 67, 89, 125, 134, 156, 175, 188, 250
Technology, 3, 6, 7, 9–11, 25, 28, 29, 32–34, 37, 40, 42, 45, 49, 66–71, 76, 84, 89, 96, 97, 100, 107, 120, 121, 124, 125, 137, 138, 140–142, 148, 150–152, 157, 159, 162, 167, 175, 181, 187, 188, 190, 194, 199, 201–203, 208–210, 212, 214, 216, 218, 219, 226, 227, 231, 232, 234, 238–240, 246, 248, 249
Tesla, 49, 247–249
Third Italy, 13, 15
Too-big-to-fail, 31, 48, 229
Toyota, 3, 4, 20, 25, 27, 35–37, 39, 40, 45, 50–52, 65, 68, 70, 71, 86, 89, 90, 92, 95, 96, 99, 102, 104, 106, 111, 112, 130, 135–138, 140, 143, 144, 148, 159, 161, 162, 165, 167, 172, 185, 186, 199, 201, 218, 219, 225, 230, 231, 234, 235, 240, 242, 243
Toyota's production system, 35, 45, 104, 165–167, 235, 238–240, 243
Toyotism, 148, 166
Training, 11, 16, 25, 32, 33, 99, 119, 122, 139, 141, 152, 175, 209, 212, 248
Training Road Map (TRM), 152
Transaction cost, 34, 84, 111, 191, 205
Transfer, 6, 25, 31, 34, 35, 43, 71, 110, 150–152, 154, 156, 167, 179, 198, 199, 201, 202, 213, 216, 218, 219, 226, 237, 238, 240, 241, 243, 244, 249
Transplant, 35, 36, 217, 237, 240
Trial and error, 66, 74, 101, 120, 125, 139, 156, 201, 204, 214, 229, 234, 238

U
Uddevalla, 39
Ulsan, Ulsan plant, 139, 150, 160, 178

Union, 24, 41, 69, 71, 73, 106–108, 112, 122, 124, 140, 150, 152, 156, 162–165, 202, 213, 236, 241, 242, 245, 248, 249

V
Value Advanced Automotive Trade Zone (VAATZ), 157
Value chain, 29, 36, 76, 120, 181, 188, 190, 247
Varieties of capitalism, 41
Veblen, Thorstein, 6, 9, 11, 16, 43, 49, 77, 219, 226, 227
Vehicle Dependability Study (VDS), 92, 95
Vehicle sharing, 246
Vendor Management Inventory (VMI), 157
Vertical integration, 20, 25, 27, 29, 30, 34, 45, 172
Vertical subcontracting system, 175
Vocational qualification system, 69
Vocational training, 23, 69
Volkswagen, 76, 78, 86, 89, 101, 154, 162, 249
  Volkswagen emissions scandal, 247
Volvo, 39, 74

W
Window of opportunity, 49, 77
Work organization, 25, 35, 39–42, 96, 104, 106, 142, 147, 159, 167, 213, 240, 249
Work standards, 38–40, 127, 129, 133

Z
0.5-tier, 76, 181